PLASTICS IN AUTOMOBILES

U.S. Materials, Applications, Markets

CRC Press
Taylor & Francis Group
Boca Raton London New York

CRC Press is an imprint of the
Taylor & Francis Group, an **informa** business

Plastics in Automobiles

First published 1994 by Technomic Publishing Company. Inc.

Published 2019 by CRC Press
Taylor & Francis Group
6000 Broken Sound Parkway NW, Suite 300
Boca Raton, FL 33487-2742

© 1994 by Taylor & Francis Group, LLC
CRC Press is an imprint of Taylor & Francis Group, an Informa business

First issued in paperback 2019

No claim to original U.S. Government works

ISBN 13: 978-0-367-44936-0 (pbk)
ISBN 13: 978-1-56676-136-9 (hbk)

Main entry under title:
Plastics in Automobiles: U.S. Materials, Applications, Markets

TABLE OF CONTENTS

i

LIST OF TABLES

I – Introduction

A. Objectives

This report is intended to provide a basic understanding of the major products, technologies, applications, marketing practices and competitive scenario of the plastic automotive business.

In addition, this study will offer current market estimates of the most important polymers broken down by specific automotive application and will furnish projections estimating the growth of each market segment over the next five years.

Inherent in these estimates and projections will be an assessment of both intraplastic competition as well as the battle between plastics and glass, rubber and metals within the automotive industry.

The impact of new technologies, volume of car production, and enforcement of federal and state legislation by the EPA will be major considerations in evaluating the automotive plastics market over the next five years.

This study will also present opportunities for new plastic systems in both current and future automotive design and applications.

1

B. Scope

Plastics usage in autos is not easily categorized. The term "plastics" denotes commodity polymers such as polyethylene (PE), polypropylene (PP), polyvinylchloride (PVC) or polystyrene (PS). However, the definition of "engineering plastics" is not quite as clear-cut. Most include nylons, polycarbonates, polyacetals, polyphenylene sulfides, polysulfones, etc. Other polymers such as fluoropolymers, polyetherether ketones, polyamides/imides, etc. are often included as well. Others in the industry refer to engineering polymers as specialty thermoplastic polymers.

The thermosetting polymers used in autos are mainly comprised of unsaturated polyesters, phenolics, polyurethanes and epoxies. These polyesters find heavy use in sheet molding compounds (SMC), and bulk molding compounds (BMC). Polyurethanes (often called urethanes) are also referred to as RIM (reaction injection molding). Furthermore, many thermoplastic and thermosetting polymers are often reinforced with glass or carbon fiber, which place them under the umbrella of "reinforced plastics", a term often used interchangeably with "composites".

Another complicating factor is the rapidly growing field of polymer alloys/blends in which two or more commodity, engineering or reinforced polymers and elastomer systems are combined. Technically, there is a difference between an alloy and a blend - the former retaining the characteristics of each component, the latter does not. This fact is often overlooked and these "mixtures" are almost always grouped together in one category.

The market has been analyzed by major polymer system and segmented into commodity, engineering and alloys/blends. The first two definitions may be somewhat arbitrary but are consistent. Splitting out polymer alloys/blends avoids double-counting and singles out this rapidly growing market segment.

Where feasible, this analysis is extended into specific auto applications in which exterior applications contain considerably more segments than interior ones.

Current and emerging technologies applicable to plastic usage in autos will be reviewed and evaluated which should enable readers to assess the impact of technical advances on the marketplace. This information will aid readers to identify opportunities which may develop for specific applications.

Market evaluations and projections will cover only domestic commercial end-use applications. However, where possible, comments on world markets and foreign competition will be included.

C. Methodology/Sources of Information

This report was initiated by a complete review of the literature pertaining to the automotive plastics market followed by exhaustive analyses of supplier trade literature. Letters were sent to about 100 suppliers of plastics to the auto industry and following review of this material, contacts were made to clarify the product position of major companies.

This information was carefully analyzed, and a preliminary market analysis was sent to the most important suppliers. Following a series of interviews, the data was reviewed, necessary changes made, and many of the key representatives within the industry were re-contacted in order to re-evaluate the results.

The base year for this study is 1993 with a forecast for 1998. As a frame of reference, market volumes are supplied for 1988, as well. All dollars will be in constant 1993 dollars and all growth rates are compounded annually.

Definitions and abbreviations will be made and indicated when first used and where considered desirable for clarity. Tradenames appearing in the report will be shown in upper-case letters.

D. Structure/Format

The SUMMARY, CONCLUSIONS and Section IV follow the Introduction. These chapters provide concise reviews of the subject matter covered in this report. Section IV covers both automotive and plastics industries.

Section V details the market information segmented by the major types of polymers: commodity, thermoplastic polyesters, thermosets, engineering polymers, polyurethanes, polymer alloys/blends, and thermoplastic elastomers. Section VI extends the polymer market analysis into the major automotive applications: interiors, exterior body, other exterior, fenders/-bumpers/fascias, under-the-hood and other.

Section VII reviews the marketing aspects (including new products) while Section VIII provides a competitive analysis of automotive plastic producers and covers product substitution and molders. The critical subject of current and future technologies is examined in Section IX, while, another vitally important aspect, Environmental/ Regulatory Considerations is carefully assessed in Section X. Of special importance are the topics of automotive plastic recycling and Corporate Average Fuel Economy (CAFE) programs.

II - SUMMARY

A. Product Markets

Automotive plastics consumption in 1993 is estimated at about 2.5 billion pounds which should increase to slightly over 2.8 billion pounds by 1998 which corresponds to an annual growth rate of 2.4% (Table 1).

Commodity polymers (made up of mostly polypropylene, PVC and HDPE), engineering polymers (mostly ABS, nylons and polycarbonate) and polyurethane account for almost 80% of total volume in 1993.

The value of the automotive plastic market is estimated at about $1.8 billion in 1993 with engineering polymers and polyurethanes accounting for almost 60% of the total (Table 2).

Thermoplastic elastomers and thermoset polymers will experience the largest growth rates while engineering polymers and polyurethanes will increase at lowest rates.

One of the reasons for very moderate growth of polyurethanes is that the largest application - auto interiors - has the lowest growth rate at 1.2% (Table 3). Interiors are predominately plastic in one form or another and growth of one polymer is usually at the expense of another.

TABLE 1

AUTOMOTIVE MARKET VOLUME BY POLYMER GROUP

(MM lbs)

	1988	1993	1998	Growth Rate 1993-1998
Commodity Polymers	800	855.6	952.7	2.2%
Engineering Polymers	545	579.6	628.9	1.7%
Polyurethanes	490	509.4	537.8	1.1%
Thermoset Polymers	182	216.8	269.5	4.4%
Thermoplastic Elastomers	113	157.0	216.0	6.6%
Alloys/Blends	109	118.6	137.0	2.9%
Thermoplastic Polyesters	55	61.4	72.9	3.5%
TOTAL	2294	2498.4	2814.8	2.4%

Source: TPC Business Research Group

TABLE 2

AUTOMOTIVE MARKET VALUE BY POLYMER GROUP

($ MM)

	1993
Engineering Polymers	601.2
Polyurethanes	441.0
Commodity Polymers	265.5
Thermoplastic Elastomers	159.5
Alloys/Blends	146.3
Thermoset Polymers	109.7
Thermoplastic Polyesters	53.5
TOTAL	$1776.7

Source: TPC Business Research Group

TABLE 3

AUTOMOTIVE POLYMER VOLUME BY APPLICATION

(MM lbs)

	1993	1998	Growth Rate 1993-1998
Interiors	1024.8	1090.0	1.2%
Under-The-Hood	386.2	439.2	2.6%
Other Exterior	315.7	368.1	3.1%
Bumpers/Fascias	234.4	286.3	4.1%
Exterior Body	185.1	227.8	4.3%
Fenders	28.3	32.2	2.6%
Miscellaneous	323.9	371.2	2.8%
TOTAL	2498.4	2814.8	2.4%

Source: TPC Business Research Group

B. Automotive Industry Overview

Automotive companies still face many problems in terms of financial pressures to cut costs and improve profits, and must respond
to environmental concerns - mainly recycling of both plastics and
metals.

Car buyers' philosophies are changing. The days when auto
dealers simply waited for customers to replace his/her auto with
the latest model are over. Most auto buyers today, have little
or no brand loyalty since they are better informed, and more
knowledgeable. In addition, prospective car buyers have broader
choices of brands and models with a multitude of options.

Consumers value dependability and will not forgive mechanical
failures - he/she is buying services even though they are still
looking for an image inseparable from comfort.

Long term prospects are a bit encouraging in that the average age
of the U.S. passenger fleet is eight years old. The number of
cars over 10 years old has increased to over 20% of the total,
many of which are candidates for replacement.

Fewer people will reach driving age in the next ten years than in
the past decade, which could be offset by the baby boomer's
generation moving into the age bracket in which people tend to
buy larger amounts of more expensive cars.

The number of passenger car models is continually increasing and
this trend will force automakers to develop shorter time cycles
for model changeovers. This strategy needs to be supported by
improved cost control procedures in order to generate profits on
lower number of units for a given model.

C. Application Markets

Table 3 detailed the markets for automotive plastics applications clearly showing that interior usage continues to be the largest segment at almost 40% of the total. Interiors are also the slowest growing segment because plastic replacement of other materials is nearly complete. The remaining plastic interior scenario is one of one polymer group replacing another.

One of the most far reaching aspects of the use and direction of automotive plastic applications is that automakers are likely to increase the number of models offered. One cost-effective strategy would base lower volume autos on a composite frame with interchangeable polymer-based body panels, fenders and bumpers.

D. Marketing

Competition among suppliers of plastics to the auto industry is more intense than in any other plastics application. All of the major suppliers have a physical presence in the Detroit area.

Most of these "area" locations include laboratories for physical, mechanical and chemical testing of materials, classroom and conference rooms and production areas complete with plastics processing equipment, e.g., injection-molding machines. The equipment is geared for processing problems-solving, tool design and productivity analysis. Some plastics suppliers have separate groups for interiors, exteriors, structural applications, etc. and/or thermoplastic and thermoset specialists.

In these laboratories, procedures are devised to test prototype parts under conditions similar to actual performance environments. Capabilities also exist for rheological, dimensional and environmental evaluations.

Plastic suppliers must provide the following in order to compete effectively: R&D expertise and support, total commitment, follow-up, responsiveness to automakers needs, etc.

Stakes are very high, and those products chosen mean a great deal to competing suppliers in terms of sales and the ability to continue profitably.

E. Environmental Considerations

Automotive engineers rank the lowering of greenhouse emissions, reducing hydrocarbon tailpipe emissions and improving fuel economy as the major environmental priorities. Most plastics suppliers view fuel economy as the number one issue, followed by hydrocarbon emissions and reducing pollution from manufacturing operations.

Both groups placed toxic waste and vehicle recyclability well back on their priority lists, although recycling receives enormous coverage.

Disposal or retread of plastic components in junked cars is a concern of automotive plastic suppliers. Several companies have established repurchasing and recycling facilities.

10

III - CONCLUSIONS

1. Automotive plastics suppliers will need to provide materials that will enable automakers to develop shorter cycles for model changeovers to meet customer preferences.

2. Automotive OEMs will continue and probably accelerate pressures on automotive plastics suppliers to meet their product specifications.

3. The recycling issue will accelerate forcing automotive plastic suppliers to provide technologies to meet environmental demands and pressures. Possible advent of federal standards would accelerate the recycling movement.

4. Intraplastic competition will intensify causing only moderate growth for most plastics with the exception of TPEs and some thermoset plastics. The concept of product substitution will become increasingly important in the coming years.

5. The most significant plastics penetration in the automotive market will be in body panels, bumpers/fascias and in selected under-the-hood applications.

11

IV Industry Overview

A. Automotive Industry

1. Historical Background

During the period 1945-1973 the worldwide auto industry had been a stable one. Powerful custom barriers protected automakers against foreign competition. The first oil crisis occurred in 1973 coupled by the Japanese launching their first offensive in North America. At this time questions were being raised about environmental safety and conservation. The second oil shock came in 1979 followed by a worldwide recession as the Japanese invasion increased.

The declining importance of the domestic automotive industry is well known. Until 1940, the U.S. produced over 80% of the world's autos, and by 1980 this ratio was down to 25%. In 1972, imports from Japan and Western Europe accounted for 15% of total U.S. car sales. Most of these imports were smaller and more economical to run and maintain.

Starting in 1975, American auto manufacturers began an accelerated down-sizing program aimed at meeting the CAFE (Corporate Average Fuel Economy) standards of 18 miles/gallon. Domestic, and foreign, manufacturers began to look for ways to reduce car weight primarily to improve gas mileage which spurred the use of plastics into this market.

German autos were made in the U.S. for a number of years, but the Japanese held out until the early 1980s. Honda began U.S. production in 1982, Nissan in 1983, United Motor Mfg. Inc. in 1984, Mazda in 1987, Toyota in 1988, followed by Diamond-Star and Subaru-Isuzu.

2. Impact of Foreign Firms, Imports & Exports

The increase, especially in the last decade, in the number of foreign firms participating in the domestic market, particularly those with local subsidiaries, has profoundly affected the entire industry.

Across-the-border transactions which involve imports, exports, joint ventures, etc. are major characteristics of the auto industry. Approximate dollar shares of U.S. imports and exports by major world geographic areas are shown in Tables 4 and 5.

A negative auto trade balance of $45-50 billion exists, which may be decreased by increased local production by foreign manufacturers.

It should also be noted that Japanese-owned companies exported autos made by their U.S. subsidiaries to Japanese, Asian and European markets.

The many cooperative ventures between the Big Three (General Motors, Ford & Chrysler) and foreign firms clearly show the complex and intertwined nature of the industry.

The following serve as examples:

--- Chevrolets are made in the U.S., Canada and Mexico
--- Chevrolet GEO cars, include compact sedans made in California via a 50-50 joint venture between GMC and Toyota
--- subcompact vehicles are made in Canada in a 50-50 joint venture between GMC and Suzuki
--- some Ford cars are built by Mazda both in Japan and Michigan
--- the Chrysler-Mitsubishi 50-50 joint ventures in Illinois builds cars sold in the U.S. by both companies.

TABLE 4

MOTOR VEHICLE & CAR BODY EXPORTS BY MAJOR GEOGRAPHICAL AREA

Canada & Mexico	64.4%
European Community	9.0
Japan	6.0
East Asia NICs	6.8
South America	0.9
Other	12.9

World Total	100.0%

Source: U.S. Industrial Outlook, 1993, Dept. of Commerce

TABLE 5

MOTOR VEHICLE & CAR BODY IMPORTS BY MAJOR GEOGRAPHICAL AREA

Canada & Mexico	38.8%
European Community	14.0
Japan	42.0
East Asia NICs	1.9
South America	0.4
Other	2.9

World Total	100.0%

Source: U.S. Industrial Outlook, 1993, Dept. of Commerce

15

The auto transplant scenario has accelerated in recent years and the following is a brief review of this aspect of the domestic automotive scene.

--- Nissan (Smyrna, TN) - began in 1983 with a capacity of 450,000 cars and light trucks by end of 1992

--- Nissan/Ford (Avon Lake, OH) - joint venture launched in mid-1992 with capacity (1993) of 125,000 minivans and 250 Econo-line van bodies

--- Toyota (Georgetown, KY) - started in 1988 and has capacity for 220,000 cars and 300,000 engines. Planning for 400,000+ cars by 1995

--- Honda (Marysville, Anna & E. Liberty, OH) - a 1982 startup with a capacity of about 500,000 cars and over 50,000 motorcycles

--- Mazda (Flat Rock, MI) - began in 1987 with capacity of almost 250,000 cars both U.S. and Japanese

--- United Motor Mfg., Inc. (UMMI)-(Fremont, CA) - began in 1984 with capacity of almost 250,000 cars - both U.S. & Japanese

--- Subaru-Isuzu (Lafayette, IN) - launched in 1989 with capacity for 170,000 cars and trucks, with anticipated 250,000 in 1995

--- Diamond-Star (Normal, IL) - began in 1988 with capacity for 250,000 cars - U.S. and Japanese

Several within the industry view the transplants as a major concern. These plants have average workers who are 10-15 years younger than in American plants which creates disproportionate health care costs in the latter.

Furthermore, it is much easier to raise capital in Japan and Germany for example than in the U.S., and a mismatch can be created unless American plants can streamline their operations.

3. Automotive Production and Sales

Table 6 provides statistics compiled by Ward's Automotive Reports and the U.S. Dept. of Commerce.

This data shows that automobile sales should increase at an annual rate of 2.7% with North American sales growing at 3.2% and Japanese imports increasing at 1.9% The differential can be accounted for by increased production of U.S. subsidiaries of Japanese firms. With the departure of Volkswagen in 1988, Japanese production and joint-ventures account for all foreign-owned, high volume auto production in the U.S.

Table 7 lists automotive production locations in the U.S. with approximate percentages of domestic production.

More recent data from Ward's Automotive Reports supports the contention that the domestic-car share of 1993 will increase with most of the gain coming from The Big Three. The transplants (U.S. subsidiary of foreign firms) - mostly all Japanese - had an "off-year" in terms of increasing the goal for additional North American capacity.

U.S. retail sales in 1992 for domestic new cars totaled 7.6 million units, an increase from 6.7 million units in 1991 (Table 8).

Table 9 lists U.S. retail sales for imported new cars for 1991 and 1992 which totaled about 2.1 million units for each year.

TABLE 6

AUTOMOTIVE SALES IN U.S. MARKET BY SOURCE

(in thousands of units)

	1990	1991	1992	1993
North American (1)	6897	6259	6762	7400
Captive Imports (2)	187	·142	155	151
Japanese Imports	1523	1341	1444	1513
European Imports	423	317	311	323
Other Imports	262	241	227	214
Total	9,292	8,300	8,899	9,601

	1994	1995	1996
North American (1)	7856	8248	8340
Captive Imports (2)	139	129	128
Japanese Imports	1571	1608	1700
European Imports	323	330	278
Other Imports	211	188	187
Total	10,100	10,503	10,633

(1) U.S. & foreign manufacturers producing in North America for
    U.S. Market

(2) Japanese made for sale by U.S. manufacturers

Source: U.S. Industrial Outlook, 1993, Department of Commerce

18

TABLE 7

AUTOMOTIVE PRODUCTION BY MANUFACTURER - 1992

	Plant Locations	Total Production	% of Total Production
Chrysler	DE, IL, MI, MO	726,466	12.0%
Ford	GA, IL, MI, MO, NJ, OH	1,377,351	22.7%
General Motors	CA, DE, GA, KS, KY, MA, MI, MO, NJ, OH, OK, TX, WI	2,653,391	43.8%
Honda	OH	435,438	7.2%
Mazda	MI	184,368	3.0%
Mitsubishi	IL (1)	148,379	2.4%
Nissan	TN	95,844	1.6%
Subaru	IN (2)	32,377	0.5%
Toyota	CA (3), TN	415,416	6.8%
Total		6,069,030	100%

(1) Joint Venture with Chrysler
(2) Joint Venture with Isuzu
(3) Joint Venture with General Motors

Sources: Plant locations: Automotive News; Production: Ward's Automotive Reports

19

TABLE 8

U.S. RETAIL SALES - DOMESTIC NEW CARS - 1991 & 1992

(in units)

	1991	1992
Chevrolet	1,100,029	1,163,250
Pontiac	490,899	555,650
Oldsmobile	446,548	516,000
Buick	541,381	626,500
Cadillac	217,402	240,400
Saturn	49,039	160,000
Total GMC	2,845,298	3,261,800
Ford	1,078,122	1,190,200
Mercury	355,886	397,500
Lincoln-Mercury	547,391	613,300
Total Ford	1,981,399	2,201,000
Plymouth	175,340	191,162
Chrysler-Plymouth	294,860	317,166
Dodge	257,575	296,821
Jeep Eagle	49,528	53,100
Total Chrysler	777,303	858,249
Honda	477,134	507,000
Toyota	347,958	367,640
Mazda	80,457	96,500
Mitsubishi	68,428	105,000
Nissan	117,034	119,000
Hyundai	20,691	22,000
Subaru	44,791	54,200
Suzuki	575	200
Total Domestic Cars	6,761,068	7,592,589

Source: Ward's Automotive News

TABLE 9

U.S. RETAIL SALES - IMPORTED NEW CARS - 1991 & 1992

(in units)

	1991	1992
Alfa Romeo	3,850	3,300
BMW	54,945	56,450
Chrysler	66,863	63,800
Daihatsu	6,826	6,900
Ford	56,463	54,000
GMC	129,049	119,500
Honda	343,650	347,300
Hyundai	105,239	110,200
Isuzu	12,909	13,150
Jaguar	12,027	9,350
Mazda	145,479	167,300
Mercedes	66,345	69,880
Mitsubishi	84,774	89,700
Nissan	299,253	307,660
Peugeot	4,119	750
Porsche	5,661	5,450
Saab	26,367	23,970
Sterling	3,522	150
Subaru	63,037	61,700
Suzuki	4,403	5,700
Toyota	407,585	435,600
VWA	114,150	105,785
Volvo	73,632	70,480
Yugo	2,932	2,800
Total	2,093,080	2,130,875

Source: Ward's Automotive News

21

4. New Programs/Design Changes

It is expected that materials and the methods used to make cars will change as the auto industry moves toward the end of the 1990s. Most automotive engineers feel that plastics suppliers who aim for the most compatible applications for their products - rather than trying to be primary replacement for steel - will be the most successful in placing their plastics into new models.

The three major "new" structural design concepts expected to be on the scene in the later 1990s will have different impacts on plastics usage, and include:

---   unitized bodies - steel "unibodies" consolidating large structural segments into molded composite parts which simulate the steel frame - but at reduced weights. A strong possibility would be SRIM parts.
---   spaceframes - "bird-cage" structures (steel or reinforced composites) containing non-structural body panels, allowing car exteriors to be quickly and radically changed without altering the basic structure. Retooling costs would also be lowered with plastic exteriors. Although the concept was pioneered with thermoset SMCs, thermoplastics may also benefit because of their greater impact-resistance.
---   monocoque shells - similar to an insect's exo-skeleton, this concept combines reinforcing structural members with "appearance" panels (both from reinforced plastics) with large components bonded together to form a car's exterior. Improvements in RTM process technology could move this method into high-volume models.

According to several sources, GMC favors spaceframes, Chrysler is leaning toward unitized bodies, while Ford is "somewhere in between".

Automakers are also rethinking their production methods which provide an impetus for plastics on car exteriors. Detroit is expected, during the 1990s, to complete its planned shift from 500,000 vehicles/year model runs to several lower volumes (50,000-100,00 cars/year) models with constantly changing "looks". Savings in tooling costs, and rapid design turnaround via plastics could outweigh the advantages of lower-cost steel stampings as model-run volumes drop.

However, several key issues regarding plastic body panels need to be resolved:

--- establishing a processing base for molding of large body panels
--- will consumers accept car exteriors that bend when leaned upon
--- solving thermal expansion problem of thermoplastics
--- need for economical methods to recycle SMC parts
--- convincing plastics suppliers to give up the idea that their products can be used on the entire car body.

One very visible roadblock to thermoplastic body panels is the small number of independent molders with large injection units. Most of these have invested very heavily in equipment geared for SMC production.

The trade press is replete with articles covering planned programs of major auto producers along with anticipated design changes. This topic, alone, could provide the basis for an entire report on its own. This section of the PATHFINDER report will highlight this subject with important representative examples but should not be considered as a complete list of anticipated new models.

Table 10 lists selected new programs planned by GMC, Ford and Chrysler.

TABLE 10

NEW PROGRAMS PLANNED BY THE BIG THREE

---   ----------------------------

GMC   -   Buick ROADMASTER body, Cadillac SEVILLE body, Suburban
          BLAZER, REATTA body, supercharged V-6 engine, SATURN
          3-door hatchback, GEO's body car (with Toyota & Suzuki),
          NORTH STAR V-8 engines

Ford -   GRAND MARQUIS, PROBE, TAURUS, SABLE, ECONOLINE, T-BIRD,
          TEMPO/TOPAZ, AEROSTAR & FESTIVA replacements, RANGER
          truck

Chrysler - Jeep CHEROKEE, EAGLE SUMMIT, Jeep ZJ wagons, Dodge ZD
            wagons, Dodge VIPER, ACCLAIM/SPIRIT

1994 GMC - ALLANTE, W-body Buicks and Oldsmobile, J-body
             Chevrolets and Pontiacs, GEO R-body cars, SATURN WAGON,
             QUAD-IV

1994 Ford - CONTINENTAL, MUSTANG, ESCORT, ZETA 4-cyl engines.

1994 Chrysler - Jeep JJ-mini-WRANGLER, standard size pickup truck

----

1995 GM - Chevrolet CORVETTE, H-body Buicks/Oldsmobile/Pontiacs,
            B-body cars and wagons (Cadillac, Buicks, Oldsmobiles
            and Pontiacs)

1995 Ford - TAURUS/SABLE, F-Series Trucks

1995 Chrysler - LX41 Luxury car, LH21/27 Sporty midsize cars, RAM
                  vans and wagons

Source: Ward's <u>Automotive</u> <u>News</u>

24

A recent Auto Show focused on "ecologically friendly" models
along with the following specific examples:

--- Buick SCEPTRE rear-drive V-6
--- Chrysler LH cars
--- BMW sporting an electric fan
--- GMC's new SATURN wagon
--- Ford's FLARESIDE pickup truck

The Chrysler LH models along with Chrysler CONCORD and Dodge
INTREPID recently went on sale. The major external body
components of the LH will be composites - mostly fenders. The
use of DuPont's BEXLOY in the LH is Chrysler's first instance of
a thermoplastic in a major, vertical exterior body panel. The
composite fenders, which are painted on the assembly line along
with steel parts saves seven pounds per car and resulted in an
80% savings in tooling costs.

GMC is also promoting the SALSA, a plastic body five-seat hatchback entry, which can be converted to a five-seat convertible or
two-seat panel truck. The SALSA is geared for warm-weather
consumers. The vehicle's exterior is glass-reinforced epoxy,
while the interior is primarily vinyl.

The SALSA is considered an example of a "concept" car whose major
objective is to stretch the limits of design, engineering and
style.

The most current news on the electric auto scene centers on
development efforts for advanced batteries. The U.S. Advanced
Battery Consortium (USABC) is developing programs at Sandia and
Argonne National Engineering Laboratories.

The USABC, formed in January, 1991 is a partnership of GMC, Chrysler and Ford with participation from the electric utility industry through the Electric Power Research Institute. By October, 1991, the consortium officially became a four-year, $260 million joint government-industry research project.

The consortium's midterm goals are to broaden the capability of electric vehicles by the mid-1990s including at least doubling vehicle ranges, and extending the life of batteries.

Electric battery systems under development include: nickel-metal hydrides, lithium polymer electrolytes, lithium (metal) sulfide, lithium alloys, lithium metal/vanadium oxide combinations.

5. Impact of CAFE and Other Environmental Issues

The Corporate Average Fuel Economy (CAFE) program has been fought by both U.S. and foreign automakers. The CAFE value, currently at 27.5 miles per gallon (mpg) sets the average for an auto producers entire passenger car fleet and is enforced by the Environmental Protection Agency (EPA).

More stringent fuel economy and exhaust-emission standards are in the offing. So far, automakers have responded with lighter cars via extended use of plastics and catalytic converters for reduced exhaust emissions.

The next stage will probably involve alternate-fuel strategies, lighter vehicle materials, new engine technologies and possibly-down the road - electric vehicles.

The drive for higher fuel efficiencies and cleaner air is often not in synch. Although burning less gasoline usually means reduced emissions of unburned hydrocarbons and carbon monoxide, oxides of nitrogen can actually increase depending on the mode of combustion.

If Congress mandates a significant increase in CAFE, larger and mid-size car production would have to decrease. In order to avoid that option, automakers are focusing on the following alternatives:

--- lean-burn combustion in which increased fuel efficiency depends on increasing the air-to-fuel ratio which can increase nitrogen oxide (NOX) emissions. New catalytic reduction techniques to counter this problem are not yet in place and federallymandated reduced NOX emissions will become effective by 1996.

--- two-stroke engines are basically lean-burn, but face the same NOX limits. All of the Big Three are investigating this option, in many cases, as joint ventures with Japanese companies. Advocates of the 2-stroke engine claim a 20% fuel economy benefit over the comparable 4-stroke engines. It is generally agreed that this option is also long-term, probably not until the end of the decade.

--- diesel engines run lean, but face the NOX problem, as well as being bulkier and heavier than 2-stroke engines. Comparatively slow acceleration and fuel odor still linger as other major diesel engine drawbacks, although past problems of smoke and noise have been mostly solved.

--- alternate fuels are another choice, but each has its own set of strengths and weaknesses. This issue has also become a political football and the winners in the natural gas/ethanol/methanol battle is still "down the road". The Big Three have adopted a neutral position. Currently there are few incentives to buy alternate fuel vehicles.

--- electric vehicles will be on the market in the late 1990s
    since California has mandated that 2% of each auto
    producer's new car fleet must be made up of zero emissions.
    No other technology can meet this requirement. All of the
    Big Three are gearing up for some electric vehicle
    production.

--- a final option might be the use of heated catalyst converters
    which would cut down on emissions during the first few
    minutes of operation - which is usually the heaviest period
    of emissions.

6. Current Marketing & Long-Term Prospects

Automotive companies face competing pressures - financial ones to
cut costs and improve profit and employers pressures to maintain
employment. Currently, the former are winning, in that lower
profit business segments are being restructured, cut back or
sold. Chrysler tried to sell its Acustar component manufacturing
company several years ago, but it was blocked by the unions.

Those subsidies that drain profits which cannot be sold will be
reduced in size. GMC has centralized global purchasing in the
U.S. following the success of these policies in Europe. GMC's
internal suppliers will suffer by these actions.

There has been an enormous volume of trade press on the recent
problems of General Motors. Turning GMC around will be difficult
since the company operates at a $4 billion annual labor cost or
about $800 per car - much higher than Ford or Chrysler.

GMC's strategy for restoring its operations to profitability include the closing of over 20 plants, laying off over 50,000 workers, making white collar workers pay more of their health costs, freezing salaries, and finally "strengthening" and streamlining products. Laying off 20,000 white collar workers and shedding noncore businesses is also on the drawing board.

Chrysler has outsourced its part logistics for Europe, and several U.S. importers have done likewise.

In regard to the implementation of many EEC policies American automakers still control most of the U.S. market, while European car manufacturers control their geographic market.

Vastly different perceptions lead to various market needs. For example, the Toyota CAMRY, assembled and sold in the U.S., is very different from the Japanese version.

Car buyers' philosophies have changed. Gone are the days when the auto dealers simply waited for the customers to replace his/her auto with the latest model. Most auto buyers today, have little or no brand loyalty because they are better informed, and more knowledgeable. In addition, the prospective car buyer has a broader choice of brands and models with a multitude of options.

The consumer values dependability and will not forgive mechanical failures - he/she is buying a service even though they are still looking for an image inseparable from comfort.

In order to capitalize on recent market-share gains, the Big Three have tempted buyers with "bargains" on many 1993 models - averaging only about 1.2% over 1992 models. The message is that 1993 models focus "more on value". Japanese car companies having poor profits raised prices on their 1993 models by over 3%.

29

The 1993 model year posed different challenges for each of the Big Three. Ford was trying to maintain its increase in share of both cars and light trucks. GMC tried to maintain its position during restructuring, while Chrysler attempted to capitalize on the success of its minivans and Jeeps with a new line of family sedans - Dodge INTREPID, Eagle VISION and CONCORDE.

The midsize sedan market is fiercely competitive with Honda's ACCORD and Ford's TAURUS - the leading sellers. Meanwhile Nissan is pushing its ALTIMA family sedan replacing the slow-selling STANZA.

Toyota will introduce several attractive small cars, as well as improving its COROLLA model. The Big Three have countered with GMC's SATURN and Chevrolet's new wagon - GEO PRIZM (GMC - Toyota joint venture). Chrysler has an "aging" group to compete, e.g., Plymouth ACCLAIM, Dodge SPIRIT and LEBARON.

Long-term prospects are somewhat encouraging for the automotive industry in that the average age of the U.S. passenger fleet is about eight years. The number of cars 12 or more years old has increased to over 20% of the total. Many of these autos are candidates for replacement over the next few years.

Although increased use of consumer leasing programs could foster long-term sales, pending fuel economy and safety legislation could increase production costs. Increased retail prices and higher operating costs could, however, restrain sales.

Several states are initiating requirements that certain percentages of new cars have improved emissions which may drive up retail prices.

It has also been determined that fewer people will reach driving age in the next ten years than in the past decade, which could be offset by the baby boomer's generation moving into the age bracket in which people tend to buy more expensive cars -- and more of them.

The number of passenger car models is continually increasing. This trend will continue and will force automakers to develop shorter time cycles for model changeovers. This strategy needs to be supported by improved cost control procedures in order to generate profits on a lower number of units for a given model.

Worldwide auto markets are maturing rapidly, which will result in even fiercer competition for market share. It is anticipated that the next five years will see more mergers, takeovers, and joint-ventures. The inevitable result will be fewer and more "evenly-matched" global companies.

Taking into account the various undefined recession scenarios ranging from - short, average, deeper, and extended - a range of 15.0-17.3 MM units have been forecast for 1996. This represents growth rates of 1.8% and 4.7% respectively. The author of this report leaves the value of such a forecast with the reader.

7. Historical Usage of Plastics

Early applications of plastics in autos included vinyl roof covers, upholstery, seat cushions, instrument panels, interior trim panels and electrical system components such as lamps and lenses. Major applications appeared during the 1970s beginning with injection molded instrument panels. Many interior items were converted to polymers such as cushioning, carpeting, trim covers/fabrics, instrument panels, dashboard components and headliners - mostly with engineering polymers except polyurethane seat cushions.

During this period, plastics made their appearance in exterior and under-the-hood auto parts. Most prominent were radiator grilles, front and rear fender extensions, body moldings, fan shrouds, fender aprons, etc. In addition, blow molded materials were introduced and RIM polyurethanes were beginning to be used in front and rear panels.

Many of these new plastics applications were driven by the need to respond to the oil supply and pricing problems of the 1970s. The auto industry undertook massive retooling in order to make the cars smaller, lighter and more fuel efficient. As a result of these efforts, the second half of the decade saw an almost 50% fuel economy improvement. The growing challenge of foreign auto makers mandated engineering and quality improvements which, to a large extent, went hand-in-hand, with the shift to plastic components.

Generally, penetration of the outer body has been accomplished with the plastic fenders and body panels. Many within the industry feel that the exterior market has been "fender and/or bumper-driven". The U.S. government's 5 MPH test standard was considered a major driving force of "getting plastics into fenders". Other auto manufacturing techniques using frames with attached plastic panels are currently underway. Plastic bumpers are already standard on many cars.

It is unlikely that new plastics will be discovered. The development of new materials with specific properties from existing polymer systems with necessary variations is one of the major goals of industry.

V - Plastic Product Markets: Analysis & Forecasts

A. Overview

Table 11 summarizes U.S. Plastics sales in 1991 & 1992 while
Table 12 cites plastic sales to major markets for 1991 and 1992.

B. Commodity Polymers

The major commodity polymers include polyethylene (PE), polypropylene (PP), polyvinyl chloride (PVC) and polystyrene (PS).

1. Polyethylenes

Polyethylene is a translucent material often fabricated into
clear, thin films. Thicker sections are more translucent than
thinner ones, and the addition of colorants and other additives
greatly enhances polyethylene's commercial importance. Polyethylenes are partially crystalline and partly amorphous and have
very wide molecular weight ranges.

Compounding of polyethylene provides specific properties for
various applications and is usually carried out by producers or
fabricators/compounders.

Within the industry, polyethylenes are usually viewed as to their
respective densities. The three major types are low-density
(LDPE), linear low-density (LLDPE) and high-density (HDPE).

HDPE can be made via the Phillips process (chromic oxide) and
results in products of very high molecular weight (40,000-
10,000,000). The molecular weight distribution of polyethylene
is very important and is responsible for many of the properties
of the polymer. HDPE, however, is mostly made via the Ziegler
process (aluminum alkyls & titanium salts) with wide variations

33

TABLE 11

U.S. PLASTICS SALES - 1991 & 1992

(MM lbs)

	1991	1992
ABS	1,125	1,285
Acrylic	625	626
Alkyd	315	320
Cellulosics	79	81
Epoxy	490	470
Nylon	556	595
Phenolic	2,663	2,957
Polyacetal	140	146
Polycarbonate	601	662
Polyester, thermoplastic	2,541	2,742
Polyester, unsaturated (a)	1,081	1,192
Polyethylene, high density	9,345	10,434
Polyethylene, low density (b)	11,492	12,307
Polyphenylene-based alloys	195	204
Polypropylene and copolymers	8,326	8,502
Polystyrene	4,895	5,197
Other styrenics (c)	1,219	1,298
Polyurethane	3,147	3,330
Polyvinyl chloride & copolymers	9,215	10,053
Other vinyls (d)	178	177
Styrene acrylonitrile (SAN)	114	112
Thermoplastic elastomers	614	655
Urea and melamine	1,658	1,737
Others	345	358
Total	60,959	65,440

(a) Resin only
(b) Includes LLDPE
(c) Excludes ABS and SAN
(d) Includes only polyvinyl butyral, polyvinyl formal and
    polyvinylidene chloride

Source: Modern Plastics

34

TABLE 12

U.S. PLASTICS SALES TO MAJOR MARKETS - 1991 & 1992

(MM lbs)

	1991	1992	% Of Total in 1992
Appliances	1,219	1,341	2.0%
Building/Construction	11,414	12,764	19.5%
Electrical/Electronics	1,945	2,166	3.3%
Transportation	2,012	2,198	3.4%
Packaging	15,467	16,540	25.2%
Furniture	1,039	1,074	1.6%
Toys	774	841	1.3%
Housewares	1,532	1,578	2.4%
Other	25,557	26,938	41.3%
Total	60,959	65,440	100%

Source: Modern Plastics

in density and branching leading to a diversity of properties.

LDPE is made via three processes: low-pressure, high-pressure tubular and a high-pressure stirred-autoclave polymerization. LLDPE, the most recent of the three is obviously the most linear (less branching) and was introduced by Phillips Petroleum. LLDPE's unique properties are used in pipe, film and wire and cable applications.

Nameplate polyethylene capacity is estimated at 25.3 billion lbs with the following producers listed in decreasing percent of the total:

Quantum	- 17.2%
Union Carbide	- 11.9%
Dow	- 11.8%
Exxon	- 9.1%
Chevron	- 8.9%
Oxychem	- 6.9%
Mobil	- 6.7%
Solvay	- 6.0%
Paxon Polymer	- 4.7%
Phillips	- 4.7%
DuPont	- 3.2%
Westlake	- 2.8%
Eastman	- 2.6%
Rexene	- 1.6%
Hoechst-Celanese	- 1.4%
Lyondell	- 0.5%

Uses for polyethylenes in autos include:

--- fluid reservoirs (washers, radiator overflows, etc.)
--- fuel tanks
--- spoilers
--- trunk linings

As an example of specific HDPE applications, Table 13 cites selective specifications of several applications for Chrysler, Ford and Honda for Phillip FORTIFLEX blow molding products.

It is estimated that 15% of U.S. cars have HDPE gas tanks, while the percentage in Europe is several times higher at 60%.

The following fuel tank material requirements are being met by HDPE products and include:

--- toughness at both high and low temperatures
--- excellent chemical resistance
--- excellent stress cracking resistance
--- unique combination of good stiffness so it does not sag under the weight of the fuel, yet flexible so it does not fail under impact conditions.
--- must be safer under crash conditions than a metal fuel tank even though HDPE will burn and not explode like some metal tanks

Reinforced HDPE has also successfully replaced steel and other metals in structural applications such as automotive seat backs and load floors.

37

TABLE 13

FORTIFLEX POLYETHYLENE BLOW MOLDING CURRENT
AUTOMOTIVE SPECIFICATIONS

Specification	Application
CHRYSLER	
AM5133	Washer Bottles, Coolant, Ducts
MS-DB53C	HVAC Ducts, Fuel Vent Tube
MS-DB471A (MS-DB24) (MS-DB47)	Washer Reservoirs Air Dist. Ducts
MS-DB471B (MS-DB439)	Coolant Reservoirs
MS-DB471C (MS-DB417)	Fuel Tanks
MS-DB471D	Fuel Tanks
FORD	
ESB-M4D145A&B	Washer and Coolant Reservoirs, Seat Backs, Consoles
ESA-M4D197-A	Fuel Tank Shields, Shock Absorber Boots
ESL-M4D197-A2	Fuel Tanks
ESA-M4D197-A3	Fuel Tanks
HONDA (MS-14-025-OP)	Headrests

Source: Phillips trade literature

2. Polypropylene

The excellent properties of isotactic polypropylene have helped
it become one of the fastest-growing commodity thermoplastics.
The polymerization of propylene is usually accomplished with
Ziegler-Natta catalysts. Catalyst performance often dictates the
process choice for the polymerization, e.g., slurry, gas-phase or
liquid-monomer.

Polypropylene may be processed via injection molding, thermoforming or extrusion. Polypropylene has outstanding combinations
of heat resistance, tensile strength, abrasion resistance,
clarity, optical gloss as well as relatively low specific density.

Polypropylene injection molded materials are used for food
containers, large boxes and tanks, while polypropylene film is
used in packaging. Stretch-extruded polypropylene-based products
have found use in textile yarns, monofilaments, straps, brushes,
etc. Extruded polypropylene pipe is utilized in industrial
waste, water and floor-heating systems.

Propylene block copolymers are important commercial products
prepared from propylene either following homopolymerization
followed by copolymerization with ethylene, propylene or by
blending polypropylene with ethylene-propylene rubber.

Nameplate capacity of polypropylene is estimated at 9.4 billion
lbs with the following producers listed in decreasing percentage
of total capacity:

Himont	–	20.9%
Amoco	–	16.4%
Exxon	–	10.5%
Fina	–	8.8%
Aristech	–	7.1%
Shell	–	5.3%
Eastman	–	5.3%
Phillips	–	4.7%
Solvay	–	4.4%
Huntsman	–	3.5%
Lyondell	–	2.9%
Quantum	–	2.9%
Epsilon	–	2.5%
Other	–	4.8%

Polypropylene has found many uses in the auto market. It can be used, without any additives, as a fender liner. Polypropylene can be effectively filled with glass fiber, but still remains somewhat troublesome to paint and maintain its gloss. Over-loading with fillers such as talc and glass can cause some breakdown of the polymer. However, notable successes have been realized with bumper beams. In this application, polypropylene is reinforced with 40% glass for compression molding.

Other polypropylene auto usage includes internal body panels, fan shrouds, and ducts, seat belts, flooring, steering wheel covers, brake and clutch cylinders, fascias, bumper liners and other interior applications. Polypropylene is also injection molded into battery components such as cases, covers and vent knobs.

Table 14 provides several examples of Phillip FORTILENE polypropylene sheet and blow molding applications.

TABLE 14

FORTILENE POLYPROPYLENE SHEET AND BLOW MOLDING
CURRENT AUTOMOTIVE SPECIFICATIONS

Specification	Application
CHRYSLER	Wood Flour Filled Sheets
DELCO DPM-4114	Blow Molded Reservoir
GENERAL MOTORS GM 6052	Air Ducts
FORD ESF M4D135	Air Ducts
FORD	Wood Flour Filled Sheets
COMPOUNDERS	TPO/Bumpers Sealants

Source: Phillips trade literature

Recent improvements in melt flow and resistance to warpage have created new opportunities in large auto parts, e.g., headlight retainers replacing ABS.

In auto interiors, polypropylene is replacing engineering resins in entire instrument panels and dashboard structures. Japanese cars use a great deal more polypropylene than ABS in auto interiors.

3. Polyvinyl Chloride

The popularity of polyvinyl chloride (PVC) stems from its high degree of chemical resistance. It also has the unique ability to mix with a wide array of additives resulting in many reproducible PVC compounds with wide ranges of physical, chemical and even biological properties. Dispersion and general purpose polymers make up most of the PVC products, although PVC latex is commercially available.

PVC is always used with additives prior to processing, one of the most important of which are heat stabilizers. Without the incorporation of these latter materials, PVC would not be a viable commercial product, because it would degrade giving off HCl.

PVC products also require plasticizers during their preparation in order to become flexible. The plasticizer must, of course, be compatible with the PVC resin. Other PVC additives in addition to heat stabilizers, include: pigments, fillers (usually calcium carbonate), colorants, flame retardants and biocides. Other optional additives are impact modifiers and processing aids.

PVC nameplate capacity is 11.5 billion lbs with the following
producers listed in decreasing percent of total capacity:

Oxychem	- 18.3%	Westlake	- 4.4%
Shintech	- 17.4%	Certainteed	- 2.3%
Formosa	- 16.4%	Union Carbide	- 1.2%
BF Goodrich	- 15.2%	Goodyear	- 1.1%
Georgia-Gulf	- 7.5%	Vygen	- 0.4%
Borden	- 7.4%	Keysor	- 0.4%
Vista	- 7.3%	Other	- 0.7%

Some of the automotive uses for PVC include: interior upholstery, floor mats, interior trim and wire harnesses, hoses and in
some anti-corrosion coatings.

Occidental Chemical's Vinyl Division (Wayne, PA) segments PVC
automotive applications as follows:

--- dispersion resins - automotive sealants
--- suspension resins - calendered goods for seats, door panels,
     and dashboards
--- suspension resins - hose applications
--- dispersion resins/suspension resins - armrests.

4. Polystyrene

Polystyrene is a high molecular weight linear polymer. The
commercial product is amorphous, transparent and stiff and can be
easily fabricated above its glass-transition temperature at which
time it becomes a viscous liquid. Acrylonitrile, butadiene and
other monomers are copolymerized with styrene to produce other
commercially important products. The incorporation of rubber
moieties into polystyrene imparts toughness which results in

43

high-impact polystyrene (HIPS). Glass-reinforcement of poly-
styrene dramatically improves its mechanical properties, strength
and stiffness.

Polystyrene foams are used for planks and boards mostly for
insulation, packaging and construction. Expandable polystyrene
is used in packaging and polystyrene foamed sheet for varied
consumer markets.

Polystyrene nameplate capacity is 6.5 billion lbs with the
following producers listed in decreasing percent of total
capacity:

Huntsman	– 19.9%
Dow	– 16.6%
Polysar	– 12.4%
Fina	– 9.9%
Mobil	– 9.7%
Arco	– 8.4%
Chevron	– 7.4%
Amoco	– 4.8%
BASF	– 2.7%
Deltech	– 2.2%
Scott Paper	– 1.9%
Karma	– 1.2%
American Polymers	– 1.1%
Dart	– 1.1%
A&E Plastics	– 0.7%

Other auto applications include: instrument clusters, fascias,
steering column shrouds, dashboard skins and radiator grilles. A
good part of the ABS auto market volume lies with engineering
polymer alloys.

5. Acrylics

Many commercial acrylic polymers are copolymers of acrylic ester monomers with one or more other monomers. These copolymerization reactions lead to the development of a wide variety of resins which are used in many different applications.

Acrylic monomers require the use of initiators to undergo bulk or solution polymerization, e.g., azo compounds and peroxides. Emulsion polymerization of acrylic esters is also an important method of preparation.

Some of the more important acrylic ester monomers include: methyl, ethyl, isopropyl, butyl, isobutyl, t-butyl and 2-ethyl-hexyl. Acrylic esters are used in coatings, paper, ceramics, caulks & sealants, films, polishes, automotive, etc.

Acrylic polymers (often called acrylates or polyacrylates) are thermoplastic and can be easily injection molded or extruded into sheet form. These acrylics have excellent non-yellowing charac-teristics, good light transmission, and are relatively heat stable. They are optically clear and have excellent impact and puncture resistance. The major disadvantage of acrylics are their poor scratch resistance. Acrylics do not have the overall strength of some engineering polymers such as polycarbonates.

Some current uses for acrylics for automotive applications include:

--- lamp housings and covers
--- instrument glazings
--- lenses
--- side marker lights and assemblies

45

Tables 15 and 16 summarize market estimates for commodity automotive polymers. The 856 million lb market in 1993 should increase to about 953 MM lbs by 1998 which corresponds to a 2.2% annual growth rate. Polypropylene, HDPE, and PVC account for almost all of the volume. The value of this market is estimated at about $266 MM with polypropylene accounting for 38% of the total.

C. Thermoplastic Polyesters (PET & PBT)

Thermoplastic polyesters are condensation products that contain ester linkages along the polymer backbone. The most important, in terms of auto usage, are poly (alkylene terephthalates). These materials which are usually reinforced (usually with glass fiber) generally have superior physical, mechanical and electrical properties. The two most important polymers in this class are poly (ethylene terephthalate) (PET) and poly (butylene terephthalate) (PBT). PET is made from ethylene glycol and dimethyl terephthalate while PBT is made similarly from 1,4-butanediol and dimethyl terephthalate. Thermoplastic polyesters can be processed in several ways such as injection molding, blow molding and extrusion (as filament, film or sheet).

PET consumption is concentrated in blow-molded bottles (50%) and extruded film (25%) and comprise 75% of the total. Engineering grades of PBT and PET usage at about 10% of total PET volume is more fragmented with transportation and electrical/electronics accounting for about 46% of the total. Major thermoplastic polyester producers include: DuPont, Eastman, Goodyear, Hoechst-Celanese, ICI Americas and Shell.

PET usage within the auto market (almost always as engineering grades) is concentrated in small exterior body panels and trim, and ignition systems.

TABLE 15

COMMODITY POLYMER AUTOMOTIVE MARKET VOLUME

(MM lbs)

	1988	1993	1998	Growth Rate 1993-1998
Polypropylene	317	342	385	2.4%
Polyvinyl Chloride	225	230	240	0.9%
HDPE	210	231	267	2.9%
Acrylics	35	39	46	3.4%
Polystyrene	5	5.0	5.0	0%
Other	8	8.6	9.7	2.4%
TOTAL	800	855.6	952.7	2.2%

Source: TPC Business Research Group

TABLE 16

COMMODITY POLYMER AUTOMOTIVE MARKET VALUE

($ MM)

	1993
Polypropylene	$101.4
Polyvinyl Chloride	67.5
HDPE	58.8
Acrylics	31.5
Polystyrene	2.3
Other	4.0
TOTAL	$265.5

Source: TPC Business Research Group

The major market for PBT in the auto industry is in large exterior parts such as fender extensions, panels and rear/side louvers. The major stumbling block of PBT polymers is warpage, which has been somewhat overcome when PBT is used in conjunction with engineering polymers as polymer alloys.

Tables 17 and 18 summarize the approximately 62 MM lbs/$54 MM thermoplastic polyester automotive market - growing at 3.5% per year.

Some smaller exterior automotive parts include fender extensions, panels and rear/side louvers. Auto bumpers remain a large potential market for PBT.

PBT resins are used to a large extent as a replacement for die-cast metal and thermoset parts under the hood. Some typical applications include:

--- distributor caps   --- ignition coil bobbins
--- rotor and housings   --- vacuum valves
--- air blower deflectors   --- transmission parts
--- deflector extensions   --- spark control housings

Some other PBT auto electrical uses are fuse blocks, lamp socket inserts, rectifier bridges, windshield motor housings and switch housings. Some additional PBT auto mechanical applications are: windshield wipers, brake system parts, window/door hardware and windshield washer switches.

48

TABLE 17

THERMOPLASTIC POLYESTER AUTOMOTIVE MARKET VOLUME

(MM lbs)

	1988	1993	1998	Growth Rate 1993-1998
PBT	45	50.2	59.5	3.5%
PET	10	11.2	13.4	3.6%
	-----	------	------	------
TOTAL	55	61.4	72.9	3.5%

Source: TPC Business Research Group

TABLE 18

THERMOPLASTIC POLYESTER AUTOMOTIVE MARKET VALUE

($ MM)

	1993
PBT	$45.0
PET	8.5

TOTAL	$53.5

Source: TPC Business Research Group

49

D. Thermosets

1. Unsaturated Polyesters

Thermoset polyesters, also called unsaturated polyesters are
made, in situ, from unsaturated acids or anhydrides and
polyhydric alcohols (e.g., glycols). The most commercially
important of these esters, especially those used in autos, are
reinforced primarily with glass fiber. Thermoset polyesters may
be cast, molded, laminated, etc., are very heat resistant, but
have limitations in terms of processing difficulties and longer
cycle times than thermoplastics.

Major thermoset polyester producers are: Ashland Chemical, Owens
Corning Fiberglass, Reichhold Chemical and Cook Composites &
Polymers.

Almost all of the thermoset polyesters used in the auto industry
are sheet molding compounds (SMC) and, to a lesser extent, bulk
molding compounds (BMC), and thick molding compounds (TMC).

SMC is a "ready-to-mold" composition of resin, glass fibers and
filler. Thermoset polyesters are the predominant resin used to
make SMC. Many prefer this material over steel because of its
cost-performance advantages. SMC can consolidate several parts
into a single, molded unit -- are lightweight, but strong enough
to withstand road stresses.

BMC is another type of composite molding material made from
resin, filler, reinforcing fibers with unsaturated polyesters,
again, being the dominant resin.

SMC is chopped strand or mat which is not randomly displaced
glass fiber as is BMC. The latter, as a result, is not as strong
as SMC.

50

TMCs are combinations of BMC and SMC technologies where high filler loading can be used with resin-unsaturated polyesters. According to proponents of TMC, it provides the lower cost of BMC with the high strength.

Automotive applications for SMC are predominantly for body panels such as hoods, roofs, lift gates, quarter panels, sliding and hinged doors. BMCs are also used in automotive exterior body panels and more recently for headlamp housings. Automotive for TMCs began with molded grille opening panels.

The 11-member company SMC Automotive Alliance (Bloomfield Hills, MI) part of the Society of the Plastics Industry, has issued its 1992 report detailing where sheet molding compound is being used in cars and trucks.

Some highlights from the report are:

--- new bumper beam applications on Cadillac SEVILLE and El DORADO and Pontiac BONNEVILLE.
--- exterior body panels by Ford Motor
--- SMC and steel cargo box and SMC rear fenders on Ford's FLARESIDE pickup truck
--- SMC sunroof for the SATURN and SATURN wagon roof panel and lift gate
--- SMC hoods for Ford's ECONOLINE van
--- SMC hoods for one version of Lincoln's MARK VII.

Tables 19 and 20 show that the thermoset polymers automotive market is estimated at 217 MM lbs which will increase to 270 MM lbs by 1998 accounting for a 4.5% annual growth rate. Sheet molding compounds are driving this market. Polyesters comprise over 80% of thermoset polymer volume. The value of this market is about $110 MM.

2. Phenolics

Phenolics are an important group of thermoset polymers mainly composed of phenols and formaldehyde. Alkyl-substituted phenols, diphenols and bisphenol A are other important phenolic starting materials.

Phenolics are used in wood composites, fiber-bonding (laminates), coatings and adhesives. Similar to thermoset polyesters, phenolics are often used with reinforcing materials, notably glass fiber.

Phenolics are relatively heat resistant, have good impact-strength and are inexpensive, but they are difficult to injection mold and are usually dark in color. Current auto applications are in "non-visible" areas such as transmission components, starter commutators, some motor housings and disk brake pistons.

More recently, engine manifolds from phenolics appeared on 1993 Ford RANGER trucks, and if accepted, parts may appear on Ford vans and truck models by the mid 1990s. Phenolics replace aluminum in this automotive application with a weight savings of almost 30%.

Major phenolic suppliers include: BP Chemicals, Durez Division (Occidental Chemical), Plaslok Corp., Resinoid Engineering and Rogers Corp.

TABLE 19

THERMOSET POLYMER AUTOMOTIVE MARKET VOLUME

(MM lbs)

	1988	1993	1998	Growth Rate 1993-1998
Polyesters	150	180	225	4.6%
Phenolics	25	27.9	33.1	3.5%
Epoxy	7	8.9	11.4	5.0%
TOTAL	182	216.8	269.5	4.5%

Source: TPC Business Research Group

TABLE 20

THERMOSET POLYMER AUTOMOTIVE MARKET VALUE

($ MM)

	1993
Polyesters	$82.5
Phenolics	18.8
Epoxy	8.4
TOTAL	$109.7

Source: TPC Business Research Group

53

3. Polyurethanes

All polyurethanes contain carbamate groups (-NHCOO-) in their polymer backbone. These groups are also called urethanes and are most commonly the reaction products of diisocyanates (or polyisocyanates) with organic polyhydroxy compounds - called polyols. The polyols can either be based on polyesters or polyethers.

Actually the term "polyurethane" is used as the general name for that portion of the chemical industry which produces or uses polyisocyanates. However, "polyurethane" is exclusively applied to the final polymeric products within that segment of the industry.

Branched (or cross-linked) thermoset polyurethanes can be obtained with higher functional monomers (polymeric isocyanates) in addition to the linear thermoplastic polyurethanes from difunctional monomers.

Polyureas are made by substituting amine-terminated polyethers for hydroxy-terminated ones used in standard polyurethanes. The resultant product contains an MDI-based isocyanate on one side and a blend of an amine-terminated polyether and aromatic diamine chain extended on the other. The use of amine-terminated polyethers strongly affects the physical properties and processing characteristics of the formulation and may also affect reactivity. These polyurea polymers exhibit improved thermal stability over polyurethane polymers.

Linear polymers have limited thermal stability, but have good impact-strength and physical properties as well as excellent processability. Urethane polymers can also be formed by the trimerization of part of the isocyanate group yielding so-called

54

"network polymers" used in the formation of rigid foams. The choice of building blocks, molecular structure and foam density determine the physical properties of polyurethanes.

The most widely used isocyanates in polyurethane manufacture are toluene diisocyanate (TDI) and a polymeric isocyanate made from an aniline-formaldehyde-based polyamine (PMDI). An important co-product of the latter is 4,4'-methylenebis-(phenyl isocyanate) [MDI].

Since the chemistry of polyurethanes is so complex, the choice of isocyanates varied, and the types of active hydrogen compounds almost limitless, new technologies and products are put forth at a more rapid rate than most other polymer systems.

At first, polyester polyols were the most commonly used material, but the less expensive polyether polyols are currently the preferred raw material. Most flexible foams are based on polyethers and have excellent cushioning properties and remain flexible over wide temperature ranges. Furniture and bedding are the largest outlet for flexible foams, while rigid foam is used primarily for insulation in the form of boards or laminates. Research is currently underway with other polymeric MDI and polyisocyanate foams.

The major automotive polyurethane application is seat cushioning. Other standard applications for polyurethanes in autos include: interior trim, glove compartments, in-situ foaming behind crash-pad film for fascia panels, rug underlay, gasketing for glazing, door panels, fenders, bumpers, vertical body panels, instrument panels, etc.

Another very important group of products is thermoset reaction-injection molding (RIM) products. Some within the industry refer to these products as polyurea RIM and they are amine-terminated.

RIM products are used extensively by the automotive industry and compete with other plastics molding processes and products, such as injection molded thermoplastics and compression molded SMC. The RIM process enables the auto industry to quickly mold large complex parts with high surface quality and enjoys significant capital advantages over other conventional large-part injection molded processes.

RIM polyurethane/polyurea products compete head-to-head with many thermoplastics and SMC materials in several automotive applications.

RIM producers cite many advantages of their products and processes. In regard to surface quality, RIM products have Class A surfaces with minimum surface porosity which is claimed to be an advantage over SMC, while the latter, according to RIM producers, is a problem with engineering thermoplastics. Paint solvent attack is considered minimal with RIM products, as well. RIM producers also point out that their materials allow for thick cross-sections with uniformly consistent mechanical properties. This allows little sacrifice in cycle time and mold design is quicker and easier.

RIM technology has evolved through several stages. The first RIM polyurethane polymers were made from polyols, glycol chain extenders and MDI and were used for fascias. The next development resulted in hybrid polymers of polyurethane-polyurea.

These materials reduce cure time and defects due to air entrapment; however, losses occur in productivity due to the spraying and cleaning of the external release agents from RIM mold surfaces.

A third stage in RIM technology shortened the cycle time by reducing the time needed to spray the mold. This technique is often called Internal Mold Release (IMR) and was pioneered by Dow Chemical.

Advanced RIM, for example, includes systems such as poly-cyclopentadiene, polyurea, nylon, SRIM and epoxy which, according to producers, provide improved structural properties and more economical processing than SMC, injection molded thermoplastics, and stampable thermoplastic sheet.

Other RIM innovations include use of inmold painting, use of reinforcing materials, upgrading of RIM elastomers, and improved automation which have increased production rates. Inmold painting is in its early commercialization stages for modular windows, steering wheels, headrests, etc. Color matching, however, could prove a problem for this technique.

Lower-cost reinforcements (milled glass) are being challenged by minerals (wollastonite) and fibrous calcium silicate especially in RRIM fenders for minivans. Surface-treated mica reinforcements are also being investigated.

Continuing innovations are narrowing the efficiency gap between RIM and standard injection molding techniques and RIM producers envision larger clamp tonnage presses and metering machines for higher throughput. It is expected that RIM parts of 100 lbs or more will be commonplace within a relatively short time.

More recently, injection-molded RIM products are being used as exterior body panels from reinforced polyurethanes and yield Class A surfaces.

A great deal has been written on the use of chloroflurocarbons (CFCS) as blowing agents for polyurethane foams. The move to replace CFCS is well underway and reliable reports indicate that CFC blowing agents in flexible foams have been halved and could be completely eliminated by 1994 or 1995. Progress with rigid foams is not as rapid.

Another serious problem which could affect polyurethane ability to compete with other systems is recyclability. More will be said of this issue in a later section of this report.

Major domestic polyurethane suppliers include: BASF, Miles, Dow Chemical, DuPont, ICI Americas and Union Carbide.

Polyurethane consumption in autos is estimated at about 510 MM lbs in 1993 growing slightly at 1.1% per year to reach about 538 MM lbs by 1998 (Table 21). Most of this volume is in flexible foam seating whose growth is flat. The good growth of RIM/SRIM is not of sufficient volume to significantly affect the total polyurethane market.

The value of the polyurethane automotive market is estimated at $440 MM (Table 22).

4. Sheet Molding Compounds (SMC)

SMCs have been discussed in several parts of this report, but this section summarizes its impact on the automotive industry - which may prove helpful to those readers especially interested in this market.

TABLE 21

POLYURETHANE AUTOMOTIVE MARKET VOLUME

(MM lbs)

	1988	1993	1998	Growth Rate 1993-1998
Flexible Foams	300	300	300	0%
RIM/SRIM	120	133	157	3.4%
Rigid Foam	60	65.6	69.3	1.1%
Other	10	10.8	11.5	1.3%
	------	------	-------	------
TOTAL	490	509.4	537.8	1.1%

Source: TPC Business Research Group

TABLE 22

POLYURETHANE AUTOMOTIVE MARKET VALUE

($ MM)

	1993
Flexible Foam	$270
RIM/SRIM	108
Rigid Foam	54
Other	9

TOTAL	$441

Source: TPC Business Research Group

The major reasons for the continued popularity of SMC in the automotive industry according to the SMC Automotive Alliance are:

--- light weight   --- styling advantages
--- parts consolidation   --- corrosion resistance
--- dimensional stability   --- ease of assembly

SMCs were introduced in the early 1970s, and by 1993 the automotive industry will use over 1.5 million panels.

Some of the recent SMC applications include:

--- Cadillac SEVILLE and EL DORADO, along with Pontiac
     BONNEVILLE, selected front and rear bumper beams
--- SATURN uses an SMC sunroof on sedans and as a liftgate on its
     station wagon
--- Ford's ECONOLINE Van employs SMC hoods with over 200,000
     units being planned for production

Resin transfer molding (RTM) is a low volume alternative to SMC, used mainly with heavy-duty trucks. The cost difference between SMC and RTM can be substantial at volumes of 5,000 units a year. RTM also requires special painting systems and a separate primer. RTM was used to make inner hoods on the new Dodge VIPER.

The SMC Automotive Alliance, an activity of the Society of the Plastics Industry's Composite Institute, is an association of 30 of the largest suppliers and molders of SMC to the auto industry. Its current members are shown on the next page.

Allied Signal Inc.	J.M. Huber Corporation
Amoco Chemical	Lord Corporation
Aristech Chemical	Menasha Corporation
Ashland Chemical Inc.	Molded Fiber Glass Companies
Budd Plastics Division	Morton International
The Complax Corporation	Owens-Corning Fiberglas
Cook Composites and Polymers	Plasticolors, Inc.
Crown Fiberglass	PPG Industries, Inc.
Dow Chemical	Premix/E.M.S. Inc.
Dubois USA	Reichhold Chemicals, Inc.
Du Pont Canada Inc.	Rockwell International
Eagle-Picher Plastics	Synthetics Products Company
GenCorp Automotive	Trans Plastics, Inc.
Goodyear Tire & Rubber Company	Union Carbide
Interplastic Corporation	Vetrotex CertainTeed Corporation

The Alliance publishes an annual list of different SMC components on domestic and imported passenger and truck lines. The 1993 list includes almost 250 SMC components, on over 100 vehicle types produced by over 20 manufacturers. Table 23 lists SMC components in 1993 Auto models. Table 24 lists selected vehicles which contain SMC components.

E. Engineering Polymers

A group of polymers designed specifically for engineering applications was introduced during the 1950s and 1960s and includes nylons, polycarbonates, polyacetals, polyphenylene oxides, polyphenylene sulfides, etc. Almost 20 years ago several new polymers were marketed which further extended the chemical, physical and thermal properties of the engineering polymer market -- polysulfones, polyetheretherketones (PEEK), polyamides/-imides, etc. Many within the industry also consider fluoropolymers as "engineering polymers".

TABLE 23

SMC COMPONENTS IN 1993 AUTOMOTIVE MODELS

Grille Opening Panel
Rear Taillight
Headlamp Doors
Cowl

Rocker Arm Covers
Front Fender
Rear Bumper
Rear Bumper Beams

Spoiler
Belt Line Moldings
Rear End Panel
Rear End Extension

Front Bumper Beams
Fender Skirt
Hood Louver
Hood

Rear Hatch
Door
Front Underbody
Lower Body Panel

Roof
Door Sills
Rear Quarter Extension
Full Filler Door

Rear Quarter Headlamp
Liftgate

Sunroof
Front Bumper

Source: SMC Alliance

TABLE 24

CARS & TRUCKS CONTAINING SMC COMPONENTS

CARS:		
	Type	Model
Chrysler	Chrysler	FIFTH AVENUE, IMPERIAL, LEBARON, NEW YORKER
	Dodge	DAYTONA, DYNASTY, SHADOW
	Plymouth	SUNDANCE
Ford	Ford	CROWN VICTORIA, MUSTANG, PROBE, TAURUS, TEMPO
	Lincoln/ Mercury	CONTINENTAL, GRAND MARQUIS, MARK VII, TOWN CAR

TABLE 24 (continued)

	Type	Model
GMC	Buick	CENTURY, LESABRE, REGALN
	Cadillac	BROUGHAM, EL DORADO, FLEET-WOOD, SEVILLE
	Chevrolet	BERETTA, CAMARO, CAVALIER, CORSICA, CORVETTE, LUMINA
	Oldsmobile	CIERA, CUTLASS, TORONADO
	Pontiac	BONNEVILLE, FIREBIRD, GRAND PRIX, SUNBIRD
	Saturn	SATURN
Imports/Transplants	BMW	BMW
	Honda	ACCORD
	Mazda	MIATA, MX-6

TRUCKS:

Chrysler	Chrysler	CARAVAN, VOYAGER, TOWN & COUNTRY
	Dodge	PICKUP, RAMCHARGER, RAM VAN
	Jeep	COMANCHE, JEEP XJ & YJ
Ford	Ford	AEROSTAR, BRONCO, ECONOLINE, F-150 FLARESIDE
GMC	Chevrolet	APVS-TRANS SPORT, LUMINA, SILHOUETTE, BLAZER, CK PICKUP, VANDURA
Imports/Transplants	Diahatsu	ROCKY
	Isuzu	RODEO
	Mazda	VAN
	Suzuki	Tire cover

Source: SMC Alliance

Engineering polymers are clearly distinguished from the high volume/lower priced commodity polymers such as HDPE, LDPE, PVC, polypropylene, polystyrene, etc.

Ordinarily, engineering polymers have superior physical, mechanical and electrical properties than their commodity counterparts. Some of these characteristics include tensile strength, stiffness, compression and shear strength, and impact-resistance. In addition, engineering polymers retain their superior properties over wide ranges of conditions, e.g., wide-temperature ranges, corrosive chemicals.

Engineering polymers are also thermoplastic which enables them to be processed by several techniques such as injection and, blow molding as well as by extrusion.

The automotive industry has, and will continue to be a significant market for engineering polymers. This interest was fueled by several production and appearance problems posed by thermoset polymers -- some of which have been corrected.

Engineering polymers can produce Class A surfaces, ready for painting, which eliminates almost all touch-up and repair operations associated with other plastics. This has resulted in new styling freedom which results in fine, molded details. Cycle times are shorter than those for SMC and RIM. Molded parts also gain their ultimate properties when they are unmolded - compared to reactive thermosets which require some aging and conditioning. SMC and RIM processes have made great strides in reducing cycle time; however, the use of molded engineering polymers has resulted in basic manufacturing economies due to immediate part usage and reproducible quality and detail. These features have

64

lead to increased robotic handling and parts assembly. This is likely to be done at optimum points in the assembly process after operations inside the car and engine compartment are completed.

Engineering polymer producers have found it more efficient and better business to modify existing product lines to match the requirements of end-users. Engineering polymer users will not pay higher prices for a material that is over-qualified for a specific application. The aim of modifying an existing product line is to produce the lowest cost material that will fit the specifications of the end-user.

1. Nylons

Chemically, nylons are polyamides produced from dibasic acids and amines via a condensation reaction. Nylons have numerical designations referring to the number of carbon atoms in its amide links. For example, Nylon - 6/6 is the product from the reaction of hexamethylenediamine with adipic acid, each of which contains six carbon atoms. The polymerization of caprolactam results in Nylon-6 via ring opening and conversion to a linear polymer. Other nylon analogs which are considered specialty polymers include 6/9, 6/10, 6/12 and 10/12 - the number of carbons of the diamine is given first.

The difference in the number of carbon atoms between the amide groups results in significant differences in mechanical and physical properties. Nylons are noted for their high strength, toughness, excellent wear properties and chemical resistance. Dimensional stability is nylon's major weakness because it easily absorbs water which results in reduction of its tensile strength and stiffness while increasing elongation by acting as a plasticizer.

However, as its moisture content rises, significant increases occur in impact-strength and energy absorbing properties of the polymer. Nylon is basically a cream-white translucent material which is nearly transparent in thin layers, which can be colored by the addition of pigments.

Reinforced nylons based on mineral and/or glass fibers, provide significant improvements in physical properties. Glass reinforcing fibers increase tensile strength; provide greater dimensional stability; improve creep resistance; yield higher stiffness and higher heat-distortion temperatures.

Mineral-reinforced nylons result in polymers with higher rigidity and higher heat deflection temperature than glass-reinforced grades. Mineral nylon reinforcement results in lower cost and lower warpage materials, but when compared to glass reinforced grades, they have lower strength, stiffness, stability, and heat resistance.

Mineral/glass blends provide an improvement over straight mineral or glass formulations. In a blend, rigidity is increased (by about 25%) while temperature resistance is improved (by about 20%).

Nylon, like many other engineering polymers, is compatible with a variety of polymers which opens up new markets in polymers, alloys or blends for automotive usage, which will be discussed in a later section of this report.

Some of nylon's automotive applications include:

--- electrical switches and assemblies
--- radiator fans
--- door handles
--- clutch components
--- mirror housings

--- radiator end tanks

--- replacements for die-cast zinc

--- truck fuel reservoirs

Overall nylon domestic nameplate capacity is about 760 MM lbs.
with DuPont, Allied-Signal and Monsanto being major producers.
Other major producers include: Elf Atochem North America, BASF,
EMS-American Grilon, Hoechst-Celanese, Bemis and Nylon Corp.

2.   Polycarbonates

Polycarbonates were introduced in the 1950s and are made by
reaction of bisphenol A with diphenyl carbonate. Polycarbonate's
key properties include: outstanding impact-strength, toughness,
transparency, high-heat resistance, dimensional stability, good
electrical properties and high-gloss appearance.

Polycarbonate's major disadvantages are poor resistance to
scratching, abrasion and chemicals, susceptibility to stress
cracking and being adversely affected by hydrocarbon solvents.

The versatility of polycarbonates has resulted in applications
ranging from transparency-dependent uses such as glazing, clear
water bottles to compact discs, computer housings, automotive
applications, and telecommunications equipment. Polycarbonates
have evolved from metal and/or glass replacement products to be
very competitive products in many applications.

Polycarbonates have become very popular with automakers
especially in combination with other polymers as alloys. Some of
the major applications of polycarbonates in the automotive sector
are:

--- instrument panels

--- head lamp components

--- decorative trim

--- seat belt parts

--- roof tops

--- grilles

--- lenses, windows

The worldwide capacity scenario has become more complex because of increases in capacity recently announced - some of which did not take place. The complications involve Dow Chemical European (80 MM lb plant in Germany) plus joint ventures in Japan with Sumitomo for 1994 completion at 40,000 tons - although currently on hold while GE Plastics now also has joint ventures in Japan. GE Plastics, which has a 150 MM lb plant in Holland, announced that it will build a plant in Japan by the end of 1993 (25,000 tons with Mitsui Petrochemical by 1994). GE Plastic's plans for a polycarbonate plant in Spain have been scaled back. The company also added 120 MM lbs at its Burkville, AL plant, while Miles increased its Baytown, TX capacity from 120 MM lbs. to 185 MM lbs.

Domestic capacity is about 820 MM lbs, with GE Plastics the leader with 500 MM lbs; Miles at 200 MM lbs and Dow with 120 MM lbs.

3. ABS

Acrylonitrile-butadiene-styrene polymers (ABS) and styrene-acrylonitrile copolymers (SAN) were developed in the late 1950s and have superior properties of toughness, rigidity and resistance to chemicals.

The resulting rubber and thermoplastic composite produced when butadiene rubber is grafted with SAN is known as ABS. The amount of rubber in ABS can vary from 5%-30% by weight, the remainder

68

being SAN copolymer. Physical properties of ABS polymers vary with their composition, and to a lesser degree, with their method of manufacture. Higher impact-strengths are achieved with increased rubber content. However, the more butadiene rubber that is added, the lower will be other physicals such as tensile strength, hardness deflection temperature, elongation, specific gravity and coefficient of expansion. There are many variations of these physical property trade-offs.

Specialty grades of ABS can be produced by blending with other polymers such as polycarbonates, nylons, PVC, styrene-maleic anhydride (SMA), etc.

Emulsion, suspension and bulk polymerization processes are used for production of ABS polymers. Each has its own set of advantages and disadvantages. The dry polymer can be made either by the emulsion or suspension process and is usually compounded into pellets prior to being sold to a plastics processor. SAN copolymers are manufactured by emulsion, suspension and continuous-mass processes. A significant portion of SAN copolymers are incorporated into production of ABS.

ABS is an attractive alternative for more expensive transparent polymer systems since it has the requisite processability, dimensional stability and solvent moldability necessary for small intricate parts. SAN's high tensile strength and modulus limits its use in applications where impact-strength is critical; however, its chemical resistance, good molding property and high-melt permit its use in many engineering parts.

ABS can be electroplated for grille applications and can be spray painted. Several semi-flexible grades are used in bumper applications. ABS, however, has relatively low heat resistance and has a tendency toward fading and/or chalking.

Additional ABS auto applications include:

--- instrument clusters
--- fascias
--- radiator grilles
--- steering column shrouds
--- dashboard skins

Significant portions of automotive ABS usage centers around alloying/blending with other polymers-notably polycarbonates.

ABS's main competition in auto applications comes from polypropylene. One ABS problem is its rigidity which results in a "tinny" sound not found in polypropylene.

The colorability of ABS is a plus factor as painting operations have declined. Painting can account for a third to half of the cost of a part. Unpainted interior parts are expected to reach 50% by the year 2000, up from 10% in 1985.

The major domestic ABS producers (capacity 2.0 BB lbs.) include:

--- GE Plastics   - (43%)
--- Monsanto   - (33%)
--- Dow Chemical - (23%)
--- Other   - (1%)

4. Polyacetals

Polyacetals are highly crystalline polymers based on formaldehyde polymerization technology. The original homopolymers, made by polymerizing formaldehyde and capping the polymer with acetate end-groups (via acetic anhydride), were introduced about 30 years ago. Copolymers, commercialized in the 1960s, were made by

70

polymerizing trioxane (formaldehyde trimer) with comonomers such as ethylene oxide or 1,3-dioxolane, which adds carbon-carbon bonds to the polymer system. These oxymethylene/oxyethylene linkages increase the copolymer's thermal and oxidative stability as well as increase its resistance to acids. The ethylene oxide comonomer provides hydroxyethyl terminal groups which increase the copolymer's resistance to alkaline environments.

The first commercial polyacetal was the homopolymer DELRIN (DuPont). An improved copolymer version, CELCON was introduced by Celanese Engineering Resins. The major volume of polyacetals sold today, worldwide, are the copolymer resins. Several years ago, DuPont introduced an improved homopolymer grade.

Polyacetals have excellent low friction properties and wear resistance and along with toughness; abrasion resistance; and moisture, heat and solvent resistance have made them preferred materials for a wide variety of applications. Polyacetals have several disadvantages in that they are affected by steam, hot water, and strong acids or bases.

Polyacetals can be fabricated as extrusions, injection moldings, blow moldings, rotomoldings, or as machined or stamped parts. Most polyacetals are fabricated by injection moldings and extrusions. Glass-and-mineral-filled grades of polyacetals are available which improve impact-strength while elastomer copolymer grades are used in applications that require ductility.

Within the automotive market, polyacetals are used for small parts which require wear characteristics such as gears, cams, pump impellers, bushings, etc. Polyacetals have also found wide application in door and window handles, seat belt buckles, switches, window-winding mechanisms, carburetor parts, etc.

DuPont, Hoechst-Celanese and BASF are the major domestic polyacetal suppliers.

5. Polyphenylene Sulfide (PPS)

Polyphenylene sulfides (PPS) are highly crystalline aromatic polymers prepared from p-dichlorobenzene and sodium sulfide. PPS, introduced in the 1970s, has outstanding high temperature stability and chemical resistance.

PPS polymers can either be injected- or compression-molded and are used in composites and applied as coatings. PPS polymers are also available in reinforced grades -- either with glass or minerals.

The unique wide varieties of melt viscosities allow PPS to be used in large, thick-walled mechanical parts as well as in small electrical components. Most PPS are processed by injection molding.

Some of the major applications of PPS resins in the automotive market include:

--- radiator end tanks
--- wheel covers
--- grilles
--- instrument panels
--- windshield wiper parts
--- vacuum-formed seats

At one time, Phillips Chemical was the sole domestic PPS supplier with its RYTON product line. In 1987, however, Hoechst-Celanese commercialized its FORTRAN line which is based on material pro-duced in Japan by Kureha and compounded in the U.S. Miles also had several PPS products based on polymers made by its parent,

Bayer, in West Germany. General Electric joined the PPS market with SUPEC, a 40% glass product made and compounded in Japan under a licensing agreement with Tosoh Steel.

In 1986, Phillips had a PPS joint-venture with Toray (Japan) and with Ciba-Geigy in Europe. Late in the 3rd quarter of 1992 Phillips announced it was selling its 50% interest in Phillips Petroleum Toray, Inc. to Toray Industries. The Phillips Toray venture was established to manufacture and market PPS using Phillips technology at a plant near Nagoya, Japan. Phillips will continue to license the technology to the former joint venture company and will sell its PPS compounding business in Japan to Toray as well. Phillips will continue to function as the exclusive distributor outside Japan for PPS produced in Japan.

The sale is not expected to affect Phillips' PPS operation outside Japan, including capacity of about 16 million lb/yr for RYTON PPS at Phillips' Borger, TX plant. Hoechst-Celanese was planning to build an 8 MM lb/yr plant in the U.S.

Although Bayer and Miles have since left the PPS market - shutting down its Belgium plant, Hoechst-Celanese will apparently build a U.S. PPS plant (joint-venture with Kureha) at Wilmington, NC scheduled for completion late in 1993.

6. Other Engineering Polymers

Several polymer systems are included in this category such as polysulfones, polyarylates, polyamides/imides, polyetheretherketones (PEEK) and fluoropolymers.

Polysulfone polymers are a group of amorphous and transparent aromatic materials containing the sulfone group. These polymers have excellent thermal stability and maintain their rigidity at elevated temperatures. They are resistant to acids and bases and

73

some organic solvents, such as alcohols. However, polysulfones are attacked by aromatic hydrocarbons, chlorocarbons and ketones. A further disadvantage of polysulfones is their susceptibility to UV degradation.

Polysulfones can be further strengthened with the addition of glass or mineral fillers. They also have the advantage of being able to accept plating and can be foamed to provide lighter weight, higher strength products. Polysulfones are available in transparent and opaque colors in both molding and extrusion grades.

Union Carbide introduced polysulfones in the 1960s but sold the business and technology to Amoco Chemicals in 1986. The polymer is sold under the UDEL and RADEL trade name. Amoco is the only domestic polysulfone producer with a 15-20 MM lb./year plant at Marietta, OH and is planning increases by 1994.

Polysulfone automotive applications include: transmission and electrical engine parts, other "under-the-hood" components, lamp bezels, battery caps, fuses, electrical ignition components, etc.

Polyarylates are aromatic polyesters made from bisphenol A and differing ratios of iso- and terephthalic acids (or acid chlorides). The resultant polymers are clear amorphous thermoplastics with excellent clarity, high-heat deflection temperatures, excellent dimensional stability, high impact resistance and good electrical properties. Polyarylates are available in several grades including those with glass fibers, and can be injection-foam- or blow-molded.

Polyarylates can be coextruded with PET and polycarbonates to form polymer alloys/blends as well as being painted, metal plated and hot stamped.

Automotive applications for polyarylates include: glazing panels, reflectors in head lamps, "under-the-hood" electricals, light-transmitting panels, lenses, head and tail lights, and other glazing components.

Major polyarylates producers include: Amoco (ARDEL), DuPont (ARYLON), and Hoechst-Celanese (DUREL).

Hoechst-Celanese has delayed an expansion, while Amoco and DuPont may exit the polyarylates business. The high cost $2-$4/lb along with a tendency to develop yellowish tints could hold back polyarylate growth.

Polyimides (thermoplastic or thermoset) are a group of polymers which are very stable at high temperatures.

Aromatic thermoplastic polyimides are characterized by excellent high-temperature resistance, good dielectric properties, and high-resistance to deformation under heavy loads.

Thermoset polyimides have good mechanical and electrical properties and are resistant to most chemicals except dilute alkalis and concentrated inorganic acids.

Polyetherimides, introduced in the 1980s, are amorphous thermoplastics with exceptional physical properties and processing characteristics. Polyetherimide polymers are characterized by high-heat and chemical resistance, good dielectric strength and transparency.

Polyamides-imides are opaque thermoplastics prepared from trimelletic anhydride and aromatic diamines. Polyamides-imides have excellent dimensional stability, creep and impact-resistance

and superior mechanical properties at very high temperatures. Polyamides-imides can be extrusion-, injection- or compression-molded.

Polyamide-imides are used exclusively in "under-the-hood" applications such as piston rings, and in the transmission and combustion chamber. Major producers are: DuPont (KAPTON and VESPEL), Amoco (TORLON) and GE Plastics (ULTEM).

Polyetheretherketones (PEEK), and polyketones find limited use in "under-the-hood" applications (e.g., fans and bearings). The former are supplied by ICI Americas (VICTREX) and BASF (ULTRAPEK) and the latter by Amoco (KADEL).

Fluoropolymers are outstanding materials as long as excessive loadbearing is not a requirement. Resistance to very high and low temperatures, excellent electrical properties, and a low coefficient of friction are the three most important fluoropolymer attributes. Two major drawbacks of fluoropolymers are their difficulty to be injection molded and relatively high prices. Fluoropolymers have, however, found some automotive markets such as:

--- brake sensors and pads
--- gaskets
--- couplings
--- seals
--- bearings

Major fluoropolymer suppliers include: Ausimont, DuPont, Allied-Signal, 3M and Elf Atochem North Americas. Daikin plans to build 5 MM lb plant in Decatur, AL (joint venture with 3M).

In addition to the use of fluoropolymers and steel overbraid in auto fuel lines, fluoropolymers find use in lines/hoses and in anti-lock brakes. Filled polytetrafluoroethylene (PTFE) have been competing with traditional elastomers, especially in seals.

Tables 25 and 26 summarize engineering automotive polymer consumption which totalled 580 MM lbs in 1993 and will grow very modestly at 1.7% to reach 629 MM lbs in 1998. ABS, nylon and polycarbonates dominate this market comprising over 90% of the total.

The value of the market is slightly over $600 MM with the same three polymers dominating this market segment.

F. Polymer Alloys/Blends

The major driving force to develop polymer alloys/blends is economic. Modifying a current product is obviously less expensive than developing a new one. Of equal importance is the shortened time frame necessary to bring the product to market. Development time is also reduced because the producer is familiar with the materials it is working with -- a shallow rather than a steep learning curve.

As an example, General Electric has a breadth of experience with polycarbonates, and the company's blending of polycarbonates with other polymers did not require additional major investments in production facilities.

Furthermore, R&D costs are lower, and from a marketing aspect, the use of polymer alloys/blends allows suppliers to enter a market more quickly.

TABLE 25

ENGINEERING POLYMER AUTOMOTIVE MARKET VOLUME

(MM lbs)

	1988	1993	1998	Growth Rate 1993-1998
ABS	275	280	287	0.5%
Nylons	175	191	215	2.4%
Polycarbonates	65	75	88	3.2%
Polyacetals	27	30	34.7	2.9%
Polysulfones	1	1.2	1.5	4.6%
PPS	1	1.3	1.5	2.9%
Other	1	1.1	1.2	1.8%
TOTAL	545	579.6	628.9	1.7%

Source: TPC Business Research Group

TABLE 26

ENGINEERING POLYMER AUTOMOTIVE MARKET VALUE

($ MM)

	1993
ABS	$233.8
Nylons	227.5
Polycarbonates	97.5
Polyacetals	35.1
Polysulfones	4.0
PPS	2.0
Other	1.3
TOTAL	$601.2

Source: TPC Business Research Group

There is a difference between a polymer alloy and a polymer blend. The former are single-phase mixtures of different polymers with some compatibility. Alloys have properties that are usually superior to those of either component considered separately - a phenomenon known as synergism. The resultant alloys usually result in tighter and denser molecular structures and are immiscible. The mixing is usually accomplished by a special processing method, or an extra component often called a "compatibilizer". Alloys generally are more resistant to heat, solvents, and ultraviolet radiation than most polymers used by themselves.

Blends are miscible mixtures of polymers, with the final product having an "average" of the properties of each component. Yet many blends exhibit synergistic properties and use "compatibilizers".

Although there are many possible combinations, there are only a few commercially important alloys/blends that have significant automotive applications.

--- polycarbonate/ABS - Miles' BAYBLEND, GE Plastic's CYCOLOY and
      PROLOY and Dow Chemical's PULSE
--- polycarbonate/PBT - General Electric's XENOY
--- polyphenylene oxide/high-impact polystyrene (PPO/HIPS) which
      is General Electric's NORYL

A fourth alloy/blend, PPO/nylon, is emerging as an important product - initially as General Electric's NORYL GTX. It is expected that other PPO/nylon products such as GE Plastics' PREVEX and BASF's ULTRANYL will be a factor as well.

There are very small amounts of other alloys/blends used in auto applications, e.g., ABS/nylon and PPO/PBT.

Borg-Warner (prior to acquisition by GE Plastics) had the original PC/ABS entry with CYCOLOY and a "true" alloy which was "compatibilized" into a blend and trade named PROLOY. The latter product has higher heat and impact-resistance when compared to the alloy version - CYCOLOY. Dow Chemical's recent entry, PULSE, is also a blend - and not an alloy.

Polyphenylene oxide (PPO) became commercially available as a base polymer in the early 1960s, however, it was not successful because it was very difficult to process. Only when PPO was blended with a high-impact grade of polystyrene did this product become accepted. One of the most important PPOs is poly (2,6-dimethyl-1,4-phenylene oxide). These materials are also known by the acronyms PPE(polyphenylether) and MPPO (modified polyphenylene oxide).

The alloyed polymer was the forerunner of most alloy/blend technology in use today. The product gained broad market acceptance as General Electric's NORYL, the most successful of all polymer alloys/blends. The amount of polystyrene in the modified polyphenylene oxide blend is generally a function of cost, not properties. The primary composition has a 50:50 ratio of high-impact polystyrene to polyphenylene oxide.

Modified polyphenylene oxide polymers are low-cost polymers, having high impact-strength and the lowest moisture absorption rate among thermoplastic engineering polymers. These properties combine a relatively low polymer cost with an ability to provide an attractive high quality finish on molded parts.

Automotive applications of polymer alloys/blends are widespread. Table 27 lists selected polymer alloys/blends used in automobiles along with specific applications.

TABLE 27

SELECTED POLYMER ALLOYS/BLENDS USED IN AUTOMOBILES

Tradename	Supplier	Application
ARLOY (PC/SMA)	Arco Chemical	instrument panels, seat belt housings
BAYBLEND (PC/ABS)	Miles	instrument panels, speaker grilles, glove compartment doors, mirror housings and wheel covers
CYCOLOY/ PROLOY (PC/ABS)	GE Plastics	exterior trim, head lamp bezels, mirror housings & wheel covers
GELOY (PC/ASA)	GE Plastics	exterior trim, tail light housings
GEMAX (PPO/PBT)	GE Plastics	exterior panels
HYTREL (PBT/TPE)	DuPont	bumper mounts
MAKROBLEND (PC/PET)	Miles	mirror housings, bumpers, gear shift knobs, mounting brackets
NORYL (PPO/HIPS)	GE Plastics	instrument panels, glove box decors, speaker grilles and wheel covers
NORYL GTX (PPO/nylon)	GE Plastics	fenders, exteriors and wheel covers
PREVEX (PPO/HIPS)	GE Plastics	vertical panels
PULSE (PC/ABS)	Dow Chemical	dashboard support & instrument panels
TEXIN (PC/TPU)	Miles	bumpers and fascias

TABLE 27 (continued)

Tradename	Supplier	Application
TRIAX (ABS/nylon)	Monsanto	"under-the-hood" components, exterior body panels
ULTRABLEND (PC/PBT)	BASF	bumpers
ULTRANYL (PPO/nylon)	BASF	fenders, exteriors & wheel covers
XENOY (PC/PBT)	GE Plastics	bumpers, exteriors & body panels
ZYTEL (modified nylon)	DuPont	emergency brakes, clutches and emission cannisters

Source: TPC Business Research Group

82

A factor which may hinder growth of polymer alloys/blends is the fierce competition between existing resin systems for automotive external body parts, an area that still has potential for growth plastics. Many of the applications that were formerly captured by alloys and blends are now facing other contending resins systems. Examples include RIM polyurethanes and polyureas for external body panels with metal-like finishes. Another example is sheet molding compounds (SMC) or bulk molding compounds (BMC) for similar applications.

Polymer alloys/blends in automotive reached almost 120 MM lbs in 1993 and will increase at slightly less than 3% per year to 137 MM lbs by 1998 (Table 28). Polycarbonate and PPO make up almost the entire market. The value of this market is about $146 MM (Table 29).

G.   Thermoplastic Elastomers (TPEs)

Thermoplastic elastomers (TPEs) are a relatively new technology that unites the processing advantages of thermoplastics with properties of vulcanized rubber. Thermoplastic elastomers may be extruded, injection molded or blow molded in the same equipment used to fabricate thermoplastics such as polyethylene or polystyrene. Fabricated thermoplastic elastomers are now frequently specified for applications formerly reserved for vulcanized thermoset rubbers. Thermoplastic elastomers compete successfully with vulcanized rubber in most end uses because higher material costs for TPEs are offset by lower processing costs. Thermoplastic elastomers are block copolymers which include styrenics, polyolefins, polyurethanes, copolyesters, and miscellaneous compositions combined mostly with elastomers such as butadiene, isoprene, or ethylene-propylene-diene rubbers (EPDM).

TABLE 28

ALLOY/BLEND AUTOMOTIVE MARKET VOLUME

(MM lbs)

	1988	1993	1998	GrowthRate 1993-1998
Polycarbonate - Based	65	72.5	86.0	3.5%
PPO - Based	44	46.1	51.0	2.0%
TOTAL	109	118.6	137.0	2.9%

Source: TPC Business Research Group

TABLE 29

ALLOY/BLEND AUTOMOTIVE MARKET VALUE

($ MM)

	1993
Polycarbonate - Based	$92.0
PPO - Based	54.3
TOTAL	$146.3

Source: TPC Business Research Group

At room temperature the polystyrene, polyolefin, etc. end groups coalesce into hard domains and physically crosslink the elastomer chains forming a three-dimensional network. This network closely resembles the structure formed by vulcanizing conventional rubbers. When the TPE is heated, the polystyrene, polyolefin, etc. end groups melt and the liquified mass can be processed as a thermoplastic material. Cooling of the melt re-establishes the crosslinked network.

Thermoplastic elastomers are manufactured and marketed in the U.S. by domestic and foreign producers of plastics, synthetic rubbers and, in the case of thermoplastic olefins, by domestic plastics compounders and include: Exxon, DuPont, Shell, Eastman, General Electric, Monsanto, Dow and B.F. Goodrich.

TPEs have wide-ranging automotive applications which include interior and exterior body components and under-the-hood parts. TPE's are generally used to control road and engine vibrations, seal in oil and grease, seal out dirt, reduce noise, transport fluids and protect wiring.

Adding to the anticipated future demand for thermoplastic elastomers is the potential for using these products in body panel constructions. Thermoplastic copolyester elastomers are currently used in body side moldings to provide low-temperature flexibility, gasoline resistance, and on-line paintability. Design engineers are now evaluating thermoplastic copolyesters and other TPEs for the construction of vertical automobile, truck, and bus body panels.

A big advantage of TPEs in automotive applications is "ease of design" which although more costly, means fewer components need to be made and assembled which helps recyclability because of less scrap cost than other polymers.

With rigids, elastomeric properties are used to dissipate impact energy rather than allow component to flex. With vehicle panels, however, the part must flex on impact, but be able to support some structural weight without distorting.

Thermoplastic styrenic elastomers are being used as replacements for PVC in the fabrication of some automotive parts. Even though PVC is lower in cost, these TPEs having lower densities and improved strength allow the manufacturer to make thinner parts. Consequently, in many applications, thermoplastic styrenic elastomers are the most cost-effective choice.

For car interiors, manufacturers use thermoplastic styrenic elastomers to make shelf mats and some shifter boots. Under-the-hood applications include sound deadeners, sealers and electrical wire coverings. Primary auto wire covering is the most important of all of these applications.

Thermoplastic polyolefin elastomers (TPOs) are used in exterior automotive parts because of their excellent low-temperature flexibility, wide ranges of stiffnesses, good paintability, and excellent outdoor weatherability and fade resistance. Thermoplastic polyolefin elastomers may be colored by internal pigmentation or the molded part can be specially-treated to allow painting to the exact shade required by the automaker.

A major application of TPO elastomers is injection-molded automotive sight shields. A sight shield is the part attached to the car between the bumper and the body of the auto. Sight shields, made of TPO elastomers, can withstand a 2.5 mph impact at a temperature as low as -22°F and yet are rigid enough to be attached to the car with fasteners screwed into the part. At the other end of the temperature scale, sight shields made from TPO elastomers, can be painted and withstand temperatures present in the finish bake oven.

In addition to bumper ends, other exterior parts made of thermoplastic polyolefin elastomers are air dams, body side claddings and moldings, fairing pieces, fender liners, rocker panel covers, rub strips, scuff plates, stone deflectors, wheel well moldings, and valance panels.

TPOs have limited applicability in under-the-hood uses since many TPOs soften at 350°F. TPOs, therefore, must be specially formulated to withstand these conditions or be used in applications where engine temperatures are lower. Examples of parts made from TPOs, formulated to withstand high temperatures, are a wiring harness protective sleeve and an air duct which channels air from the air cleaner to the carburetor.

TPOs can be specially compounded to have superior sound-deadening properties. These materials are used to fabricate fire-wall blankets, for front-wheel-drive cars and air conditioner systems, which absorb the sound originating in these areas. Sound-deadening formulations are also used to make speaker enclosures for high-fidelity automotive sound systems.

Thermoplastic vulcanizate elastomers have gained acceptance primarily as substitute products for nitrile rubber in under-the-hood and exterior sealing applications. Thermoplastic vulcanizates have superior thermal stability and durability, are much easier to process than NBR rubbers, and offer reduced variability for molded parts. Many rack-and-pinion boots, windshield washer screens, fuel-line and wiring covers, cold-air intake tubes and exterior weather stripping are made from thermoplastic vulcanizates.

At one time, TPUs were used for bumper fascias and other related products. The Department of Transportation (DOT) downgraded automotive bumper impact specifications from 5 to 2.5 mph. TPUs have been replaced by TPOs. TPUs are still used in some exterior body parts.

Thermoplastic copolyester elastomers have replaced a great deal of reinforced rubber parts in automobiles. In addition, these TPEs are beginning to appear in painted fascia and in several under-the-hood applications as well. Examples of the latter include: air ducts, rack-and-pinion boots and wire and cable harnesses.

Thermoplastic elastomers, the most rapidly growing segment in automotive plastics reached 157 MM lbs in 1993 and will increase to 216 MM lbs in 1998 (Table 30). TPOs account for almost 60% of the total. Market value of TPEs is about $110 MM (Table 31).

H.   Other Polymers

This category includes polymer systems not covered previously such as: reinforced plastics, sheet molding compounds (SMC), bulk molding compounds (BMC), weatherable polymers, etc.

Industry technology can often become confusing. Often the term "composite" is used to describe a "reinforced plastic".

Reinforced plastics are difficult to define as a separate category since they comprise polymer systems or substrates which have been modified with a reinforcing material such as glass fiber, carbon fiber, etc. Sheet molding compounds (SMC) reinforced with glass fibers, glass-reinforced polyurethanes, or phenolics are just a few examples of "reinforced plastics". High-impact polystyrenes, and high-impact engineering polymers are others.

TABLE 30

THERMOPLASTIC ELASTOMER AUTOMOTIVE MARKET VOLUME

(MM lbs)

	1988	1993	1998	GrowthRate 1993-1998
TPOs	66	90	125	6.8%
Copolyesters	23	30	37	4.3%
TPUs	8	9	9	0%
Styrenics	2	3	3	0%
Other	14	25	42	10.0+%
TOTAL	113	157	216	6.6%

Source: TPC Business Research Group

TABLE 31

THERMOPLASTIC ELASTOMER AUTOMOTIVE MARKET VALUE

($ MM)

	1993
TPOs	$79.2
Copolyesters	46.0
TPUs	14.0
Styrenics	2.8
Other	17.5
TOTAL	$159.5

Source: TPC Business Research Group

89

90

VI. Application Markets: Analysis & Forecasts

A. Overview

A brief history of plastic usage in the automotive field was described in an earlier section of this report. Generally, automotive applications require materials that maintain high-strength over wide temperature ranges. Exterior winter temperatures can reach - 40°F while upper-end temperatures are in the 200°F-300°F for disk brakes covers and some under-the-hood applications.

Automakers are likely to increase the number of models offered. One cost-effective strategy would base lower volume autos on a composite frame with interchangeable polymer-based body panels, fenders and bumpers.

The following summarizes those automotive areas where plastics have had the largest impact - the remainder of Section VI will assess the potential for further growth in each of these applications.

--- bumper systems - from steel/aluminum to plastic, steel
    beams have covers made from elastomer-modified polypropylene
    or polyurethanes
--- panels (front & rear) - from steel panels, zinc die-cast
    grilles to SMC grilles and polyurethane fascias and from
    glass to plastic headlamps
--- body panels - from steel/aluminum to composite materials
--- door panels - from steel to polyesters (SMC)
--- engine - from steel to engineering and reinforced commodity
    polymers
--- chassis - from steel/aluminum to plastics for gas cap covers,
    fuel tanks and liquid filler housings
--- seats & floors - blow molded HDPE seat backs and cushion
    frames and reinforced polypropylene seat buckets

91

--- instrument panels/trim - panels mostly converted from a
   metal to plastic with soft crash pads, injection molded door
   and covers, luggage compartment liners
--- under-the-hood - electrical housings, wirings, heater &
   air-conditioning housings, cooling fluid overflow tanks,
   window washer containers, etc.
--- fenders - RIM polyurethanes and reinforce thermoplastics
   slowly replacing steel

One of the earliest mass produced cars to use thermoplastic body
panels was the Honda CIVIC CRX. The panels were molded of
thermoplastic alloys with ABS, polycarbonates and other
materials, while the front and rear bumpers were molded with
modified polypropylene.

Plastic obviously does not rust like steel and since car buyers
are keeping their vehicles for longer periods than in the 1970s
and early 1980s, this advantage assumes more and more importance.
According to plastics suppliers, the higher cost of the plastic,
compared to the cost of the steel, is more than offset by reduced
capital investment for tooling, economies in fabrication and
Assembly, more efficient development and reduced labor costs.

Plastic parts are usually one-third cheaper to produce than metal
parts in low-volume applications. The advantage dissipates,
however, when volume reaches about 500,000 parts/year because
steel tooling is more durable than plastic equipment.

B. Interiors

Interior applications include instrument panels, seats,
upholstery, dashboards, and buttons/switches/knobs and trim.

Intense intra-polymer competition exists for interior applications: ABS, polypropylene, polycarbonates, alloys/blends, compression molded polyesters and polyurethanes. These polymers
must also compete with non-polymeric materials such as steel,
wood fiber and fiberboard.

In regard to interiors, the decision factors influencing
polymer/material choices might include the following: final per
piece cost on a per unit size basis, dimensional stability,
surface bondability, in-mold colorability, part processability,
acoustics, aesthetics, etc.

The standard instrument panel uses a foam-padded sheet steel
frame, the vertical surface contains the openings where the
instrument clusters, glove boxes, clocks and radios are fitted.
Plastic sheets have replaced most steel sheets. Many autos now
use an engineering polymer or corresponding alloy for a one piece
instrument panel.

ABS and polycarbonates (mostly as alloys/blends) are the most
widely used for instrument panels. Both these materials are
excellent insulators, highly impact-resistant, have high temperature distortion resistance and can be easily molded into a
variety of shapes.

93

There are several types of dashboard designs. One uses a steel frame wrapped with polyurethane foam and covered with an ABS sheet. This technique is adequate for most applications, since steel provides rigidity, the foam supplies the cushioning effect and ABS affords the toughness, abrasion and temperature resistance required.

Newer designs are composed of a single dashboard made of engineering polymers such as PPO/high-impact polystyrene alloys, eliminating the need for steel as well as the foam and ABS overlay. These dashboards are impact and temperature resistant and offer weight reduction over previous ones, however, they are more expensive than the steel/foam dashboards.

Many fabric interiors are made of PVC, which also is the polymer choice for roof liners, door panels and other related interior surfaces. PVC is relatively easy to decorate and maintain. Seat cushions are often made of polyurethane foam, overlayed with a PVC fabric.

The following lists several interior applications along with the polymer most often used:

--- seat frames, traditionally made of steel, are now using more and more engineering polymers
--- head liners - polyurethane foam, covered with PVC
--- door and side panels - mostly PVC, polyurethane foams
--- carpeting - acrylics and floor mats primarily calendered PVC
--- window winders - traditionally made of steel and die cast zinc, being replaced by plastics especially nylon and polyacetals
--- heating/air conditioning knobs and steering column covers - nylon
--- IP substrates - ABS, PC, SMA, PC/ABS blends, PP
--- IP surfaces and seat fillings - polyurethanes foams

94

Polypropylene (mineral or talc-filled) especially in Japanese cars, has replaced a great deal of ABS and ABS/polycarbonate interior usage particularly in trim applications, e.g., interior door panels, speaker grilles.

The new front-and-side-impact regulations (to increase occupant safety) have focused attention on instrument panels (IPs).

In regard to side safety impact laws, by September, 1993, 10% of cars produced must adhere to this legislation. This will increase to 25% by 1994, 40% by 1995 and 100% by 1996 and these regulations apply to both 2- and 4-door cars.

Obviously, this is an opportunity for more energy absorbing materials in doors and foamed plastics are probably the best way to obtain energy absorption without adding excessive weight.

Some of these materials could include polystyrene foam (on low end) to more costly foamed polypropylene and polyurethane foams.

A standard product for the IP retainer (main substructure) is glass-reinforced styrene-maleic anhydride (SMA) but ABS/ polycarbonate PPO/HIPS and polypropylene have become strong competitors. SMA is more brittle than ABS/PC and PPO/HIPS, but there are IP designs that employ foam pads or skins which make it less important if material fails ductally. In IPS, the polyurethane foam is sandwiched between a retainer (carrier) or substrate frame and a "skin" on top usually made from PVC, but TPEs are moving in. IPS now evolve into full-fledged structural members of autos, although a single large structure component is doubtful.

95

The new airbag regulations are activated with about 100 lbs. pressure, so IPs must not only hold the airbag but also keep it in place while it functions. Clearly, airbag requirements will be factored into IP designs.

Several polymer systems are vying for the knee bolster market - a necessary piece of equipment for preventing passengers from slipping under the airbag. Polyurethane foams, PULSE, Azdel's composites, GE's PPO alloys and impact-modified ABS/PC are being used. Polypropylene could also become a factor in this market.

One likely outcome of side-impact regulations will be hybrids of metals with engineering polymers for door modules with RIM polyurethane foams on the surface.

The popular mini-vans are using more ductile polyurethane foams to replace brittle door panel materials such as wood fiber composites with polypropylene binders.

The side-impact regulations have auto makers looking at polyure-thane bolsters on door panels which, on impact, would dislodge from the panel to protect passengers from chest to mid-section.

These same regulations will also have impact on door structure material so that they may become more complex because there is less mass to ensure occupant safety with side impact than with frontal collision. Are side door airbags a possible solution?

New design and aerodynamic features have caused an increase in overall glass surface and coupled with downsizing results in more heat in less space in automotive interiors. Often called "greenhouse on wheels", there is a need for interior materials that hold up under heavier solar levels and are more UV stable.

In the final analysis, intraplastic competition is strongest within the interior segment of the automotive plastics market. In summary, several overall factors will be influenced in determining the future of this market.

--- increasing demand for a larger percentage of recyclable automotive parts
--- developing of polyolefin catalysts which will result in significant upgrading of these polymers impacting engineering resin consumption
--- improved methods of producing laminated components in a single-step, automated process to reduce costs
--- increased usage of passive safety devices (airbags) that increase structural demands on IPs, door panels and steering wheels

Other developments to watch:

--- polyolefin-cushioned IP with glass-mat thermoplastic carrier covered with semi-flexible skin with sheet foam backing
--- will SRIM show that parts consolidation and load bearing benefits outweigh cost
--- modified styrenic IP (PC/ABS, modified PPO, glass-filled ABS or SMA) with softskin ASA or TPE
--- engineering polymer blow-molding (in its early stages) needs to improve structural tolerances for panel uppers but will probably not be able to meet IP retainer requirements
--- all polyurethane IP made of SRIM and foam cushion molded in place

Dow Plastics recently unveiled an all plastic instrument panel to be used in 1994 cars. The IP uses neither steel nor glass reinforcements, and is cheaper and lighter than current IPs, according to Dow.

European, Japanese and U.S. automakers appear interested as are engineering resin suppliers since these new IPs contain more engineering polymers than previous versions.

These lighter "metal-free" IPs will be needed to counter additional car weights due to the addition of air bag modules and knee bolsters.

Flexible polyurethane foams and ABS accounted for the largest part of automotive interior polymer usage (48%) of the total of slightly over one billion lbs (Table 32). PC-based alloys, polycarbonates, HDPE and polypropylene will exhibit the largest growth rates.

C. Exterior Body

Steel and sheet molding compounds (SMCs) are still the mainstays for flat-panel exterior body applications. Horizontal panels have always been more of a problem for plastics substitution than vertical ones because the former must be able to withstand sagging at elevated temperatures. This is the main reason why volume compression molded SMCs have, so far, been the major polymer system used. Critical factors material/polymers choices include: Class A Finish and final per piece cost per unit size, physical strength properties, impact-strength, in-line painting capability, weight, corrosion resistance, etc.

In addition to SMCs, other thermoset resin systems that are available for body panel applications include: RIM, SRIM (polyurethane, polyesters), RRIM, pultrusions (vinyl and polyesters).

Thermoplastics have the additional problem of not having sufficient injection presses to mold many of the horizontal parts.

TABLE 32

AUTOMOTIVE INTERIOR POLYMER MARKET VOLUME

(MM lbs)

	1993	1998	Growth Rate 1993-1998
Flexible Foam	300	300	0%
ABS	196	201	0.5%
HDPE	138	160.2	3.0%
Polyvinyl Chloride	115	120	0.9%
Polypropylene	109.5	123.2	2.4%
Rigid Foam	65.6	69.3	1.1%
Polycarbonate	52.5	61.6	3.2%
Polyacetals	22.5	26.0	2.9%
PPO - Based	16.2	17.9	2.0%
PC - Based	7.3	8.6	3.3%
Thermoplastic Styrenics	2.2	2.2	0%
TOTAL	1024.8	1090.0	1.3%

Source: TPC Business Research Group

The evolving TECHNOPOLYMER stampable RTP sheet technology does produce parts with Class-A surfaces. This process does eliminate secondary finishing operations as well as recycling, but will these factors be enough to offset the higher direct costs? Even advocates of stampable sheet technology for horizontal applications concede that "full-acceptance" might not take place until the mid-1990s.

Competition is intense within the polyurethane area where "horizontal" RIM processes are claimed to be faster than the conventional RIM which could increase the market for polyureas in vertical panels. New generations of fast-cycling polyurea formulations are driving this market segment. Engineering polymers would be the loser in this situation. Doors, on the other hand, present more of a problem for thermoplastics because dimensional stability is more critical. SMCs are often favored for designing door systems. Inner structures of doors can be welded steel skeletons, encapsulated with PUR.

In sum, the exterior body market (often called panels) is shared by three major groups of polymers and technologies:

--- injection-molded engineering polymers and TPEs
--- RIM polyurethanes/polyureas
--- compression molded SMCs (and some resin transfer molded
       polyester)

Other systems with lesser volumes include: ABS, nylons, PC/ABS and PPO/nylon.

SMC growth has been fueled by its lighter weight, ease of styling and lower tooling costs. It is generally agreed that SMCs have greater dimensional stability than most thermoplastics and can be painted on-line much like steel.

SMCs major deterrent is a poor recycling profile but it can be pyrolyzed or reground and used as filler.

Plastic-bodied cars have been around for years, and include: Pontiac's FIERO, General Motors Minivans, Pontiac TRANS SPORT, Chevrolet's LUMINA, Oldsmobile's SILHOUETTE, CAMARO, FIREBIRD, VIPER, GMC's SATURN etc. The latter has received enormous tradepress with its "hang-on" panels of thermoplastic alloys/blends, and front and rear ends of modified TPEs.

GMC's SATURN examined several options for horizontal TPEs and vertical exterior body panels which included: SMC (polyester), RTM (urethane and polyester, RIM/urethane and polyurea), and injection-molded alloys. The requirements were for low-cost, high-production run products with Class A surfaces and in-line paintability.

The choices were: door (PC/ABS); fascias (TPOs); and front fenders (PPO alloys).

Rationale for this extended use of injection-molded thermoplastics includes: reclaim/reuse of scrap, design flexibility, corrosion resistance, impact-strength, high durability and lower weight.

Injection-molded thermoplastics are becoming more acceptable for vertical panels, but still have a long way to go, e.g., large horizontal panels such as hoods, roofs and rear decks. Dimensional stability and moisture-induced warpage being the two most prominent problems. It may well be that processes other than injection- molding may be needed to surmount these disadvantages.

Although reinforced unsaturated polyethers still dominate large horizontal exterior body panels, most of these applications are compression molded SMC. Lower volume panels, however, are being made more often with resin transfer molded reinforced plastics.

General Motors redesigned 1993 F-body cars such as the CAMARO and FIREBIRD and have door outer panels, roofs and rear deck molded of SMC. In-mold coating (SMC) techniques have allowed SMC parts to match or exceed the mirror surface obtainable with sheet metal. SMC processing has a longer molding cycle which is a deterrent to more promising potential.

Most recently, an SMC hood was being used on Ford's latest luxury car, LINCOLN MARK VIII. SMC has been used on Ford hood's before, on ECONOLINE and AEROSTAR vans which have smaller, uncontoured areas not likely to wobble or show painting irregularities.

Many observers feel that penetration of plastics into body panels has not occurred at the rate envisioned a decade ago. Steel-makers have reacted by introducing double-sided galvanizing to resist corrosion, etc. Although over 10% of automotive panels may be converted from steel to plastic by the mid 1990s, steel is still economical for large runs and many plastics have limits in regard to paintability.

Thermoset polyesters are the major polymers used in the exterior body automotive segment which is estimated at 185 MM lbs in 1993 and will increase to 228 MM lbs in 1998 (Table 33).

D. Other Exterior

This application encompasses exteriors excluding body panels, bumpers, and fenders.

Grilles are often viewed as part of the trim and molding market, and many are made from ABS which can be electroplated to achieve the "chrome" look. Several "luxury" models use grilles made from

TABLE 33

EXTERIOR BODY AUTOMOTIVE POLYMER MARKET VOLUME

(MM lbs)

	1993	1998	Growth Rate 1993-1998
Thermoset Polyesters	135.0	168.7	4.5%
Polyvinyl Chloride	11.5	12.0	0.9%
Polypropylene	10.3	11.5	2.2%
RIM/SRIM	13.3	15.7	3.4%
TPOs	9.0	12.5	6.8%
Copolyester TPEs	6.0	7.4	4.2%
TOTAL	185.1	227.8	4.2%

Source: TPC Business Research Group

polycarbonates and PBT. Some feel that polycarbonate grilles look more like a metallic grille than those made from ABS. Plastic grilles are also thought to be easier to mold than metal grilles.

Currently, RIM urethanes are widely used in painted automotive fascias such as air dams, spoilers, side moldings, and grilles. Thermoplastic copolyester elastomers (TPEs) are just now moving into this application, based on superior strength and an excellent surface for painting or for sputter chrome coating. All usage is still relatively low, but the potential for these materials is bright.

Thermoplastic elastomers are slowly replacing EPDM vulcanized rubber in exterior body seal applications such as weather stripping, trunk lid and hatchback window seals, tailgate lock seals and stop light gaskets. In dynamic sealing applications, TPOs are preferred. For static applications such as weather-stripping, tail light and vent window seals, thermoplastic styrenics are usually recommended.

As fabricating technology advances and new varieties of TPEs are introduced with improved compression set and lower stiffness greater quantities of vulcanized EPDM will be displaced from automotive primary seal applications, such as door and hood-to-cowl seals. In some potential sealing applications TPEs cannot be used because of high glass-to-rubber friction.

The conversion to all-thermoplastic headlamps by the automakers has not been uniform. Polycarbonates are the current plastic of choice for most auto lighting applications such as headlamps, reflectors, etc., while the use of polyarylate reflectors is increasing.

104

Copolyester TPEs with high impact-strength have been used on trim cladding on several cars. RIM urethanes, which have been the predominant material used in flexible fascias, may face some competition from these injection molded polyesters.

Improved aerodynamics are being designed into autos resulting in less circulation around brakes for cooling. As a result, temperature-resistant thermoplastics are being used for wheel covers, which must withstand mechanical stress from flying road debris. To meet these requirements, glass and mineral-reinforced nylons and alloys such as PPO/polystyrene and polycarbonate/ABS have become very popular for this application. Plastic wheel-covers are at least half the weight of aluminum, die-cast zinc and stainless steel types.

Several recent product applications of note:

--- nylon 6 chosen for automobile wheel covers for the 1991 Volkswagen JETTA GL Sedan. This 30% mineral reinforced material was supplied by Miles, Inc. (DURETHAN)

--- plastic headlight lenses have been approved in Europe. This segment had previously been an exclusive glass market to polycarbonates and other clear thermoplastics. This will provide opportunities for both Japanese and U.S. automakers who export to Europe. Thermoplastic headlight lenses have been used both in the U.S. and Europe for some time but a strong glass lobby blocked their usage on the European market. The first European car with plastic headlight lenses should be on the road in 1994.

--- Ford's luxury car, LINCOLN MARK VIII contains RIM polyure thane grilles. The product (FLEXIBLE BRIGHT) starts as a black urethane grille and is coated with indium - a silver-white metallic element. The new process results in a two-tone effect, produces a lightweight finished part, rules out corrosion, etc.

Many polymers are used in other exterior automotive usage which totalled 315 MM lbs in 1993 which will increase at 3.2% per year to 368 MM lbs by 1998 (Table 34). ABS, thermoset polyester, acrylics and TPOs account for two-thirds of the total.

E. Fenders/Bumpers/Fascias

Most observers agree that plastic injected molded fenders are more economical than steel when volumes range from 100,000 to 150,000. Current cycle times match assembly line targets which also result in reduced part inventories when compared to steel. Sandwich molding of several types of thermoplastics is also being used. During the fender-making, resins are injected into a mold together or in sequence creating a product that can withstand oven bake temperatures with a Class-A finish.

Specific decision factors to be considered for energy absorbers, bumper beam and fascia components of the entire bumper system include: tooling/per piece costs, impact-resistance, crash tests (2.5-5 mph), design flexibility, part consolidations, in-mold colorability, etc.

The late 1980s and early 1990s saw General Motors put injected molded fenders on Buick and Oldsmobile models made primarily from PPO/nylon alloys. Substantial savings on tooling costs resulted. Late 1980 Chevrolet CORVETTE fenders were made from toughened amorphous nylon, while RIM polyurea was chosen for General Motors APV minivans.

Cadillac also chose PPO/nylon for its fenders for several early 1990 models along with front fenders for SATURN. Although alloys/blends are still the plastic material of choice for

106

TABLE 34

OTHER EXTERIOR AUTOMOTIVE POLYMER MARKET VOLUME

(MM lbs)

	1993	1998	Growth Rate 1993-1998
ABS	70.0	71.7	0.5%
Thermoset Polyesters	45.0	56.3	4.6%
Acrylics	35.1	41.4	3.4%
TPOs	31.5	43.7	6.8%
PBT	27.6	32.7	3.5%
Polycarbonates	18.7	22.0	3.3%
Copolyesters	15.0	18.5	4.3%
Polypropylene	13.7	15.4	2.4%
Nylons	13.4	15.0	2.3%
HDPE	11.5	13.3	3.0%
PPO - Based	11.5	12.8	2.2%
TPUs	9.0	9.0	0%
Polycarbonate - Based	7.2	8.6	3.6%
PET	5.0	6.0	3.7%
Polyacetals	1.5	1.7	2.5%
TOTAL	315.7	368.1	3.2%

Source: TPC Business Research Group

107

fenders, competition may come from glass/mineral reinforced PET
in the near future. Another emerging plastic fender material is
BEXLOY K-550 from DuPont.

The LESABRE and GMC's other H-body models, the BONNEVILLE and 88
switched from steel to plastic fenders for 1992 models.

Generally the plastic competitive scenarios for the components of
the entire bumper system are:

--- energy absorbers - EVA (injection-molded), polypropylene
      expanded beam foam, and polyurethane foam
--- bumper beams - PC/PBT, polypropylene (injection- and
      compression molded); vinyl esters (compression molded),
      SRIM (polyurethanes), RTM (polyurethanes and polyesters)
--- bumper cover (fascia) - PC/PBT, polypropylene (injection-
      molded), RIM (polyurethane and polyurea).

Thermoplastics used in the production of bumpers are usually made
from ABS, polycarbonate, amorphous nylon and polypropylene and
polycarbonate/PBT alloys. Most thermoplastic bumper parts are
injection molded, but some may be blow molded or compression
molded. Bumper components include bumper covers, energy
absorbers and backup beams.

Plastic bumpers are cheaper than steel or aluminum ones, and
weigh less than half of their steel counterparts. In addition,
plastic bumpers meet federally-mandated impact-resistance
requirements - bumpers must withstand a collision of five miles
per hour. The styling, flexibility (dent resistance) and elimin-
ation of corrosion have made plastics the preferred bumper
material especially when painting is not necessary.

Ford, an early entry into plastic bumpers, was in polycarbonate/PBT alloys. Modified ABS and impact-modified polycarbonates have also been employed.

There is intense competition among various plastics for the fascia market which includes RIM polyurethane/polyurea, polycarbonate/PBT (XENOY) alloys, thermoplastic elastomers and thermoplastic polyesters. More recently, reactor-modified polypropylene and copolyesters (TPEs) have challenged other polymers.

Several 1993 models have fascias made from polypropylene modified with EPDM rubber - some are painted while others are integrally colored. The latter are TPO-based, the former are mostly made up of RIM polyurethane. The TPO-based bumper may be limited to large volume model runs; but are meeting with increased acceptance. Several recent models have polymer alloy rigid bumpers; yet there is continuing interest in flexible bumpers with Ford thinking about a switch to a more elastomeric system.

Each of the competing resins has its own set of marketing advantages and disadvantages. Mid-size cars seem to prefer RIM which is paintable but the lower-end is controlled by TPOs. Several thermoplastic polyesters are challenging TPOs.

Will polymer alloys and RIM lose out to TPOs since there is pressure to move away from primers and paints? Paint adhesion may be a problem and matching the body may also cause problems for TPOs.

Higher heat-stability, low-temperature toughness and faster cycles with new polyurea systems have created interest as materials for fascias which could result in a promising polyurethane market.

One reason why polypropylene is being considered is that it can be more easily melted down and molded into other parts. This recycling technique is more easily accomplished than regrinding RIM polyurethane. Bumpers are important elements in recycling since they are relatively easy to remove.

Two potential markets exist, one is the plastic bumper and fascia market for trucks - still mostly metal. The second opportunity is with bumper beams, for the most part also metal.

XENOY and compression molded composites (polypropylene continuous glass matte alloy) are leading candidates for "plastic" bumper beams, the "so-called" long fiber reinforced thermoplastic (LFRTPs). However, RIM polyureas have superior dimensional stability and advanced technologies could propel them into a plastic bumper market. Other engineering polymers such as polycarbonates, SMAs, polyphenylene sulfides (PPS) cannot be counted out. Pultrusion and resin transfer molding (RTM) are processes that may become important in bumper beam technology.

It should be noted that bumper beams are an integral part of the structure of the auto and are mounted on hydraulic cylinders to transfer the stress of a collision from the bumper to the entire car frame.

The following are several recent noteworthy events in this automotive applications segment:

--- GMC's LESABRE and other H-body models, the BONNEVILLE and 88 switched from steel to plastic fenders for the 1993 models
--- Ford is expected to unveil a low-volume step-side model in its full-size F-series pickup line with plastic cargo box side/fender moldings similar to those on some Chevrolet and GMC sportside models

110

--- RIM urethane fascias of the front and rear bumpers are used
on Ford's luxury LINCOLN MARK VIII. This model also has
stamped-steel front and rear reinforcing beams and expanded
polypropylene foam energy-absorbing inserts
--- a new two-component epoxy primer has been developed which
increases paint adherence to TPO bumper fascia applications
by Akzo Coatings, Inc. Molders generally needed to use
adhesion promoters as well as primers before applying
topcoats

Polypropylene usage led the fender polymer market (Table 35)
while RIM/SRIM was the most dominant polymer in bumpers/fascias
(Table 36). The combined markets totalled 263 MM lbs in 1993.
Bumper/fascia growth rates are about 50% greater than the fender
market.

F. Under-The-Hood

One of the earliest plastic under-the-hood auto applications was
PVC clad copper wire enclosed in a phenolic harness. Phenolics
are also used for distributor caps, rotor caps and housings, wire
clusters and switches. PVC has replaced many harness and
cladding operations, whereas nylons and polyacetals, however, are
employed in more critical areas, especially ignition and
distributor rotors. Some electrical assemblies and connectors
are made of polycar- bonates and PBT. Batteries and battery
casings are made mostly from polypropylene, the only metal being
the tip of the electrolytic cell and top connectors.

Reinforced polypropylene and nylon are the plastics preferred for
fans, although some cooling fans are made with HDPE. Engine
cooling fans are usually composed of glass reinforced nylon.
Some models use glass reinforced polypropylene shrouds for metal
fans as well as for timing belts. Other manufacturers are
plating their fans for the "metallic" look since many consumers
expressed concern over the strength of a "plastic fan".

111

TABLE 35

FENDER POLYMER MARKET VOLUME

(MM lbs)

	1993	1998	Growth Rate 1993-1998
Polypropylene	17.0	19.3	2.9%
RIM/SRIM	6.7	7.8	3.1%
PPO - Based	4.6	5.1	2.1%
	-------	-------	-------
TOTAL	28.3	32.2	2.6%

Source: TPC Business Research Group

TABLE 36

BUMPER/FASCIA POLYMER MARKET VOLUME

(MM lbs)

	1993	1998	Growth Rate 1993-1998
RIM/SRIM	106.4	125.6	3.4%
PC - Based	58.0	68.8	3.5%
TPOs	49.5	68.8	6.8%
Polypropylene	20.5	23.1	2.4%
	--------	-------	---------
TOTAL	234.4	286.3	4.1%

Source: TPC Business Research Group

Radiator covers are often made from nylon, while HDPE is being used for timing belt covers, radiator expansion tanks, pulleys and gears, washer fluid tanks, power steering and brake fluid reservoirs, fuel tanks, and liquid reservoirs, e.g., windshield washer and radiator overflow tanks.

TPEs have made significant strides in automotive under-the-hood applications. Copolyester TPEs, because of their superior strength, thermal stability, and resistance to auto operating fluids are used to make clean air ducts, rack and pinion boots, wire and cable harnesses, etc. An area of significant potential for TPEs is in the drive train area where they are used to make engine and transaxle seals, formerly produced from nitriles or other oil-resistant vulcanized rubbers. Many TPEs have superior tensile and tear properties as well.

The evolution of smaller cars and more efficient engines, more aerodynamic styling, turbocharging and government-required emission control equipment have resulted in engine compartment temperatures in excess of 400°F. Coupled with the fact that engine fuels can now be ordinary gasoline or a blend of gasoline with oxygenated organic compounds such as methyl or ethyl alcohol or methyl-tert-butyl ether (MTBE) will make further demands on plastics that can be effectively used in under-the-hood applications. This is an opportunity for chemically-resistant plastics.

Important rationales for these engineering advances are:

--- meet demands of higher miles per gallon
--- the hotter the combustion - the cleaner the emissions

Furthermore, streamlining the front end and dropping the nose not only increases styling and aerodynamics, but reduces packing and compartment space. This results in less air circulation, more hot air and higher temperatures.

Driving forces for increased use of plastics in under-the-hood applications include: improved sealability, noise and weight reduction, overall design flexibility and potential for savings in equipment costs, ease of parts consolidation, color coding of wires and perhaps most important, longer car warranties.

Some of the simplest under-the-hood parts are hardest to reach or replace, even for hoses and ignition cables.

More recent plastic under-the-hood applications are: fuel lines, fuel rails, impeller shafts, gaskets, camshafts, valve covers and intake manifolds. In other words, plastics have replaced metals in a great many areas.

Hoechst-Celanese claims that its plastic rings perform better than metal rings. The company's CELAZOLE T (polybenzimidazole) rings may be in production within a few years.

A particularly promising area for plastics in under-the-hood applications is intake manifolds. Automakers were not sure about the idea of connecting plastic to a hot engine block. These are networks of pipes that transport air into the engine cylinders that allow fuel to burn. Currently, most are made from aluminum, but are being substituted by thermoset or thermoplastic polymers molded over a metallic core (lost-core injection molding) made of low melting alloys which is melted out of manifold after the part is injection molded.

This new system results in a more even flow of air and a more efficient engine (boosts horsepower by 3%-5%) at about half the weight of an aluminum intake manifold plus the fact that plastic intake manifolds do not get hot like aluminum. Current intake manifold polymers being employed are nylon, phenolics and PPS. General Motors claims that more than one million of its cars will have plastic intake manifolds by 1994. The company indicated that its manifolds will be injection molded unlike earlier models via lost-core injection-molding. Ford is planning plastic intake manifolds using glass fiber reinforced thermosets. The General Motors intake manifolds will be ZYTEL nylon reinforced with glass fibers, while the lost-core technology version is PPS.

Suppliers have cited advantages of each in terms of cycle time, technical capability, ease of processing and cost. The lost core version produces a one-piece manifold, the conventional injection molded one has two parts molded into one - the latter being less capital-intensive.

The primary factors behind the conversion from metal to plastic are lower costs and weight along with noise reduction.

Key plastics suppliers will be DuPont, Hoechst-Celanese, Phillips and Solvay Automotive.

Fuel rails are pressurized pipes that control flow of gasoline from the gas tank to fuel injectors. Currently, they are made of aluminum or low carbon steel. Nylon, phenolics and PPS began replacing metal in 1993 models.

Non-metallic intake manifolds had their debut with Ford's 1993 RANGER pick-up trucks - and many more will follow. Initially, General Motors will produce its plastic intake-manifolds in-house.

Although most of the attention in this application segment (development of technology and carmaker's acceptance) has been on air-intake manifolds, other rigid plastic components such as valve covers and pump housings are considered bright prospects for growth.

As engines become less mechanical and more electrical and computerized, high-performance plastics will find an increasing role as housings to protect delicate circuitry from the hazardous environment of engines exposed to high heat, road salts and chemicals such as leaking fluids. Major polymers used as housings include: nylon, phenolics, polypropylene, PVC and PBT.

Key performance criteria for under-the-hood plastics applications include: heat resistance (over 350°F); dimensional stability; chemical resistance; and noise and vibration damping.

Some of the high volume applications for under-the-hood parts such as manifolds, covers, and housings, are being injection-molded in house. GMC, for example, plans to produce almost all of its intake manifolds for V-6 cars on existing captive machinery.

An important custom molding operation focusing on the intake manifold market is Handy & Harman Automotive Group at Auburn Hills, MI. The company has developed a PPS manifold for an automaker that incorporates molded-in fuel rails.

General Motors' first completed new engine in about a decade contains a great deal of plastics -- 1993 Cadillac ALLANTE. There is no intake manifold but the fuel and air induction system contains thermoplastic tuning tubes, phenolic distribution plates, nylon fuel-rail assemblies and fuel injection systems.

Nylons and polypropylene and HDPE are the largest volume under-the-hood polymers with 78% of the total 1993 volume of 386 MM lbs (Table 37). The market segment should increase to about 440 MM lbs by 1998 corresponding to an annual 2.6% growth rate.

G. Miscellaneous Applications

The five previously discussed automotive applications cover most of the entire car, however, there are several instances where it is not entirely clear where some of the plastics are being used. In other words, having reasonable market data on the total consumption of specific plastics, it was not possible to account for this value via the five discrete segments: Interiors, Exterior Body, Other Exterior/Bumpers/Fascias and Under-The-Hood. As a result, we assigned the difference to an "Other" category. The most important application in this category is plastic fuel tanks which first appeared in the mid 1970s. In the mid 1980s, Ford's AEROSTAR minivan was the first model to use HDPE gas tanks, exclusively. By early 1991, Ford had about 15% of its models equipped with HDPE gas tanks with Chrysler and GMC planning some conversions in the early 1990s.

Environmental problems surfaced in California early in 1991, in terms of hydrocarbon (HC) emissions through the HDPE gas tanks. At that time, rules limited HC emissions to 4 grams loss over 2 hours. This has been upgraded to 2 grams/2 hours. Apparently, HDPE is more permeable than steel. The plastic industry response during 1991 was barrier treatments of HDPE via fluorination or sulfonation -- possibly as a co-extrusion process.

Air Products & Chemicals recently announced that it has developed an improved fluorine-based barrier technology which will enable plastic fuel tank manufacturers to meet more stringent emission standards. Recent tests conducted at Kautex have resulted in

TABLE 37

UNDER-THE-HOOD POLYMER MARKET VOLUME

(MM lbs)

	1993	1998	Growth Rate 1993-1998
Nylons	152.8	172.0	2.4%
Polypropylene	89.0	100.1	2.4%
HDPE	57.8	66.8	2.9%
Phenolics	22.3	26.5	3.5%
PBT	17.6	20.8	3.4%
Polyvinyl Chloride	11.5	12.0	0.9%
PPO-Based	11.5	12.7	2.0%
Copolyesters	9.0	11.1	4.3%
PET	6.2	7.4	3.6%
Polyacetals	6.0	7.0	3.1%
Polysulfones	1.0	1.2	3.7%
Thermoplastic Styrenics	0.8	0.8	0%
PPS	0.7	0.8	2.7%
	-------	-------	-------
TOTAL	386.2	439.2	2.6%

Source: TPC Business Research Group

hydrocarbon permeation rates as low as 0.1 grams/24-hours, significantly lower than those achieved with currently available barrier technology.

Kautex claims that this new technology also improves barrier performance of surface-treated polyethylene with alcohol-containing fuel blends. These AIROGUARD plastic fuel tanks offer permeation performance similar to those using multi-layer extrusion techniques while maintaining long term structural integrity of the monolayered fuel tank.

The use of the HDPE gas tank still gained favor due to its light weight and good safety features. However, the California Air Resources Board (CARB) applied pressure on the EPA to tighten its standards - for allowable limits for HC fuel vapor loss for the entire vehicle measured by industry's sealed housing for emissions (SHED) over time while fuel in the tank is heated to 24 hours.

Two issues which arose during that time were claims that use of barrier technologies resulted in losses of HDPE mechanical properties and that hydrocarbon loss also stems from fill pipe and other attachments to the fuel system.

In terms of allowable emissions, automakers usually divide it into three parts: fuel system (fuel tank and fill pipe), power train and transmission and remainder of car - each is allowed 0.6 grams/day.

Some claim that it is the mating of steel (sealed with rubber components) that yields HC loss at about 0.5 grams/day. This needs to be reduced eventually to zero grams/day.

The three North American plastic gas tank manufacturers [Ford, Solvay, Automotive and Kautex (Canada)] are confronting this issue of HC emissions. One recent development is a multi-layer blow molded tank with a nylon middle layer to provide the fuel vapor barrier - but this is a costly process.

If the proposed 1995/1996 regulations of a maximum of 2 grams of HC/24 hour is adopted, permeability of plastics will have to improve. CARB is also pushing increased use of oxygenated fuels (e.g., methanol, ethanol) in order to reduce emissions of pollutants. This complicates the issue since oxygenated materials permeate more easily through barrier plastics. Currently, barrier materials, such as ethyl vinyl alcohols appear to provide the best permeation resistance.

The near-term result of these events has been cancellation of over 250,000 plastic gas tanks by General Motors for its L-cars. Plastic gas tank manufacturers are still optimistic because of steel's inability to withstand corrosion with new types of fuels. It may be an economic production decision with plastic gas tanks being more cost effective at lower production volumes.

Other issues brought up by both steel and plastic interests are: costs of tooling, coating of steel, need of rubber hoses with steel tanks, vs. one-piece filler necks and vapor return line with plastic tanks, etc.

Nylon fuel lines have made their appearance with several General Motors' Packard models. They are flexible, convoluted tubes made in a continuous blow molding process similar to that used to make electrical conduits. The tube's flexibility allows the use of a new modular fuel reservoir assembly as well as having a high crush resistance which could be important in a collision.

In another development, polymer fiber-optic multiplex systems (which permit simultaneous transmission of multiple signals) are expected to replace conventional automotive wiring. Within a single car, as much as 200 feet of polymeric optical fiber (POF) along with insulative jacketing could be utilized. POF systems could also replace over 10 lbs of crosslinked polyethylene, PVC and other plastic wire insulation.

The rapid growth in automotive electronics is responsible for the interest in fiber-optic control. POFs are not only capable of carrying large volumes of communication data, but are easy to install and connect and reduce wiring weight by over 50%.

It is estimated that about 500,000 American cars will have limited fiber-optic data transmission control by the 1995 model year and about half of North American autos will have complete systems by the end of the decade.

The new POFs are expected not only to replace copper wire but glass optical fibers in auto data communication and illumination applications as well.

Polyvinyl chloride and polypropylene are the leading automotive polymers in the Miscellaneous category accounting for 54% of the 1993 total of 324 MM lbs (Table 38).

TABLE 38

MISCELLANEOUS AUTOMOTIVE POLYMER MARKET VOLUME

(MM lbs)

	1993	1998	GrowthRate 1993-1998
Polyvinyl Chloride	92.0	96.0	0.9%
Polypropylene	82.0	92.4	2.4%
Nylons	24.8	28.0	2.5%
HDPE	23.7	26.7	2.4%
ABS	14.0	14.3	0.4%
Epoxy	8.9	11.4	5.1%
RIM/SRIM	6.6	7.9	3.7%
Phenolics	5.6	6.6	3.3%
Polystyrene	5.0	5.0	0%
PBT	5.0	6.0	3.7%
Acrylics	3.9	4.6	3.4%
Polycarbonates	3.8	4.4	3.0%
PPO - Based	2.3	2.5	1.7%
PPS	0.6	0.7	3.1%
Polysulfones	0.2	0.3	8.4%
Other TPEs	25.0	42.0	10+%
Other Polyurethanes	10.8	11.5	1.3%
Other Commodity	8.6	9.7	2.4%
Other Eng Polymers	1.1	1.2	1.8%
TOTAL	323.9	371.2	2.8%

Source: TPC Business Research Group

VII. Marketing

A. Overview

There is intense competition among suppliers of plastics to the auto industry. These suppliers, however, are united in their battle with steel, aluminum, glass and rubber producers.

The commitment of plastics suppliers to the automotive market is clearly demonstrated by their active representation in the Detroit area. Table 39 lists selected plastic supplier centers in the Detroit area.

Plastic suppliers to the automotive industry have the following objectives especially when considering the "presence" in the Detroit area: helping to deliver cost-effective solutions, develop prototypes, fabricate new modules, and provide quality alternatives.

Most of these "area" locations include laboratories for physical, mechanical and chemical testing of materials, classrooms and conference rooms and production areas complete with plastics processing equipment, e.g., injection-molding machines. The equipment is geared for processing problem-solving, tool design and productivity analysis. Some plastics suppliers have separate groups for interiors, exteriors, structural applications, etc. and/or thermoplastic and thermoset specialists.

In these laboratories, procedures are devised to test prototype parts under conditions similar to actual performance environments. Capabilities also exist for rheological, dimensional and environmental evaluations.

TABLE 39

SELECTED PLASTIC SUPPLIER CENTERS IN THE DETROIT AREA

	Location	Product or Service	Startup
GE Plastics	Troy	Engineering Thermo-plastics	Spring '88
Budd Company Plastics Eng Design Center	Troy	Sheet Molding Cmpds	August '87
Dow Chemical Automotive Center	Southfield	Engineering Thermo-plastics, Poly-urethanes	Summer '88 (New Office)
DuPont Auto-motive Products	Troy	Engineering Thermo-plastics composites	Summer '88
General Electric Plastics, Appli-cations Dev Center	Southfield	Engineering Thermo-plastics	Spring '88
Hoechst-Celanese Automotive Dev. Center	Auburn Hills	Engineering Thermo-plastics	June '87
ICI Americas Technical Center	Sterling Heights	Polyurethanes, Composites	Spring '88
Monsanto Auto-motive Supply Center	Auburn Hills	Thermoplastics	Spring '88
Hercules	Troy	METON resins	Spring '92
D&S Plastics (1)	Auburn Hills	Injection-Molding Equipment/Paint Line	Spring '92

(1) joint venture of Dexter Corp. & Solvay Group

Source: TPC Business Research Group

124

Plastics suppliers located in the Detroit area are fully committed to the automotive industry and, in some cases, consider their presence and capabilities extensions of their engineering departments.

The recurring theme that plastics suppliers would like to leave us is their ability to respond to market needs and provide technically credible solutions from concept to production, by working with the OEMs and its molding.

Each supplier touts its design capabilities and innovative automotive applications for their products carefully pointing out that their staff brings experience from other industries.

Plastic suppliers are attuned to the needs of the auto industry and it really is a matter of "putting your money and technology on the line".

Many plastics suppliers describe how processors come to them to design and develop materials for specific applications. The automaker styles the parts, the molder has to make it, and the plastics supplier has to make it work in plastic.

One example cited in one plastics supplier's trade literature was wheelcovers. Tests are designed for all factors that affect the part's performance in the real world. Since wheelcovers must survive impact from rocks, the engineers "became authorities" on the "survival of wheelcovers" -- what makes them nick and break.

Tests were conducted on how much torque was needed to pop a cover from a wheel turning corners, fatigue and resilience. Answers were not always found by standard tests, e.g., notched-IZOD impact tests -- new ones had to be devised.

125

Prior to the application "hits-the-road", prototype parts are put through trial runs and face short- and long-term mechanical, thermal, chemical and electrical tests.

These almost remarkable series of tests, even considering that they are described in supplier literature, attest to the extensive support provided by plastics suppliers to the automotive industry.

In another instance, a large plastics supplier approached GMC engineers about an idea for redesigning a CORVETTE bumper beam. With GMC's OK, the supplier worked with a custom molder. Using sophisticated computer techniques, the supplier and molder developed a prototype beam and made a one-piece structure in place of the old two-piece box section design.

The redesigned part was lighter, saved over 10% on labor and materials and improved the fit between fascia and hood. CORVETTE, impressed with the new part, specified it for one of its later models.

These activities are obviously expensive and time consuming, but the competition is intense, the stakes very high and a long siege of missed opportunities could drive a plastics supplier out of this market.

It is generally agreed that the bulk of the displacement of steel, aluminum, glass and rubber by plastics has already taken place. The marketing battle now focuses on intra-plastic product substitution not only in terms of chemical entities such as RIM polyurethane vs. engineering polymers vs. TPEs, but whose product is chosen for a particular auto model for which year.

The stakes are very high, and those products chosen can mean a great deal to the competing suppliers in terms of sales and the ability to continue, profitably.

Reproduction without written permission of the publisher is strictly prohibited.

Plastic suppliers must provide the following in order to compete effectively: R & D expertise and support, total commitment, follow-up, responsiveness to automakers needs, etc.

Another group of companies, molders, are vitally involved in the auto market. Many of these companies supply sheet molding compound materials and are oriented toward thermoset plastics, others supply thermoplastics.

Because of this increased competition, automakers can afford to place more pressure on plastic suppliers to meet their requirements. The plastic suppliers would prefer automakers to design a car with plastics in mind.

Plastic suppliers are also fully aware that the downsizing of U.S. cars is pretty much completed so that future weight reductions may be difficult. Furthermore, automakers want to reduce the "average number" of different plastics used per car to ease the emerging recycling issue.

Steel and aluminum producers have not been inactive as their markets became eroded. Aluminum suppliers have been pushing the concept of their advanced recycling profile as a tool in their efforts to stem plastic's encroachment of their market. Steelmakers have been trying to improve metal stamping and die design.

The dismal sales of the auto industry over the past several years and the impact of imported cars (stabilized recently) along with the many "transplants" (mostly Japanese firms producing in the U.S.) are other issues directly impinging on marketing strategy and activities of plastics suppliers.

B. New Products

Very often the number and quality of new product introductions are testimony to the vitality of an industry. Even though the 5+% annual growth rate of many plastics and their automotive applications are a thing of the past, the industry is continually sparked by innovations. These advances often account for plastic substitution for metals, but, more recently, stem from intra-plastic competition.

This section will highlight significant new product introductions over the past few years in close chronological order grouped by plastic supplier.

Akzo Chemical:

--- DURASOFT, polyurethane coatings ("leatherlike" feel) for interior trim parts

Allied-Signal:

--- reinforced blow-molding grade of nylon for air-intake components

Amoco Chemical:

--- polyphthalimide resin (AMODEL) relative of nylon-- an ultra-high performance material for door and window regulator handles, hub caps, wheel covers, and emission control valve housings. These new resins are designed to bridge the performance gap between traditional engineering thermoplastics (e.g., nylons, polycarbonates, polyacetals) and high-cost specialty polymers such as PPS, polyether-imides and liquid crystal polymers.

Arco Chemical:

--- DYLARK SMAs with long fiber glass reinforcements for
    structural seat backs (Dodge VIPER)
--- ARPRO expandable bead molded foam energy absorbers in an
    attempt to make headway against injection-molded honeycombs

Ashland Chemical:

--- AROTECH vinyl ester resins for thermoset SMC and BMC
    applications, RTM polyesters able to withstand
    high-temperature paint ovens.

Dow Chemical:

--- PULSE (PC/ABS) injection-molded materials as IP retainers -
    introduced blow-molded grade and is developing SRIM retainer
    concept for its SPECTRIM polyurethane systems and SMC
    designs using DERAKANE vinyl esters
--- developed a new IP design of three major components injection
    molded from unreinforced PC/ABS. The unit will comprise the
    retainer, knee bolster and cross-car beam
--- SPECTRIM polyureas (RIM) for fascias - both filled and
    unfilled
--- SPECFLEX fully-formulated flexible molded polyurethane
    systems for interior applications including IPs, door
    panels, arm/head rests

DSM Engineering:

--- DSM developed "next generation" STAPRON (SMA copolymer) with high-heat and impact-resistance for proposed applications in front grilles, wheel covers, exterior trim and IP components

--- STANYL nylon 416 for transmission components engine and valve covers containing 50% fiberglass - the company hopes to produce larger parts in the future such as intake manifolds and valve covers

DuPont:

--- first recyclable thermoplastic composite stampable sheet system (XTC) for horizontal and vertical panels having the following advantages over SMC thermoset: melt-processable, recyclable, lighter in weight, more environmentally acceptable (SMC emits styrene fumes). Company has patent for its technology but XTC has yet to undergo field tests on a car. XTC will not give a Class-A surface, but DuPont claims minimal post-finishing is all that is required to achieve the surface-standard. However, DuPont asserts that XTC will run on existing compression molding equipment although SMC processes would have to add an air convection press. XTC also has a higher coefficient of thermal expansion which causes it to expand and contract with changes in temperature.

--- NOMEX aramid pressboard and glass composite inside exhaust shields for heat protection

--- alkylated chlorosulfonated polyethylene elastomers (ACSIUM) for timing belt materials

Exxon Chemical:

--- advanced (reactor) TPOs (EXXTRAL) as candidate to replace
vinyl in instrument and door panel application and may not
require painting. Other possibilities include bumper
fascia, body cladding and rocker panels
--- MYTEX, an engineered polypropylene developed to compete with
PC/ABS particularly with interior parts
--- TAFFEN structural thermoplastic composite which is a
compression-moldable steel of chopped glass reinforcements
in an olefinic matrix

GE Plastics:

--- for new LOMOD copolyester TPE products with Class-A finishes
for exteriors (soft bumper fascia) to compete with other
copolyesters RIM polyurethanes and painted TPOs

BF Goodrich:

--- developmental line of flux PVC composites designed to replace
painted TPUs for body-size molding and cladding trim

Hercules:

--- METTON dicyclopentadiene monomer (glass reinforced) is milled
& reaction injection molded for possible vertical exterior
panel applications. An attempt to compete with polyurethanes, polyesters and TPOs in bumper fascia market

Himont:

--- reactor-made TPOs (HIFAX) for rocker panel and vertical
cladding -- filled TPOs to compete with TPUs

Hoechst-Celanese:

--- long fiber high impact-strength polypropylene (CELSTRAND) in
    battery trays, floor pans, seat shells and IPs
--- polyarylate marker and signal lenses, reflectors, etc.
--- high-impact polyacetals for critical parts such as pump
    impellers, housings, fuel-system components, gears, slides,
    window cranks, etc.
--- polyimidizole blends in several melt-processable grades for
    piston rings, brake components, valve rings and bearings

ICI:

--- RIMLINE, a two-component polyurethane for bumper beams and
    floor pans

Miles, Inc.:

--- APEC high-temperature polycarbonates for lighting
--- MULTRANOL polyols for high-resilient polyurethane foams for
    seat cushions
--- several new reinforced nylon grades (DURETHAN) for
    applications in injection-molded door handles, mirror hous-
    ings, and wheel covers
--- MAKROLON polycarbonates for prismatic light lenses
--- BAYFLEX polyureas as fender materials for the Jeep WRANGLER -
    a first in the industry
--- BAYFILL, a foamed energy-absorber for modular bumper systems.
    The product is used with a SRIM or roll-formed aluminum beam
    with reinforced RIM fascia.

<u>Monsanto</u>:

--- TRIAX (PC/ABS) for plating applications to compete with zinc
    die cast and stainless steel for grilles, wheel covers &
    trim strip to add "distinction" to upscale models - "chrome
    look"
--- SANTOPRENE (AES) rubber-modified polypropylene for air ducts
    on Subaru LEGACY
--- VYDYNE glass reinforced nylon for radiator and tanks, engine
    fan shrouds
--- CENTREX ASA terpolymers to compete with polycarbonates in
    cowl vents

<u>Phillips</u> <u>Chemical</u>:

--- polymethylpentene (CRYSTALOR) - some of which are glass
    reinforced - for engine parts, power trains, exterior and
    interior body components, etc.

C. Relationships Between Plastics Producers & Auto Industry

The relationship between plastics producers and the auto industry
has been mentioned several times in this report, especially
earlier in this Section. However, several general points need to
be brought out.

The automotive industry specifies product and performance
requirements which may be directed to both plastics and/or
molders or compounders. Product specifications must be met for
each grade of individual plastics. The Big Three have "line
call-outs" grouped by ASTM test for each plastic product which
are tied into company specifications.

Automotive specifications are met by each product meeting certain physical, chemical and mechanical properties such as density, notched Izod impact, pressure stress, torque, etc.

Each major plastics supplier provides a list of specifications for which their products have been tested and approved.

Gaining approval for a product for a given model for a specific year is the short term goal for plastic suppliers. Without getting "a piece of the action" for a few years would probably cause a plastics supplier to withdraw from the business.

A hypothetical example might clarify a possible scenario. A plastics supplier (Allied-Signal) approaches Ford with a "new resin". Ford obtains a "property profile" for this new material and assigns it a number (ESF-M4D78-A for Allied's CAPRON nylon 8200 grade). Ford will write its product specifications only after the resin "does its job".

Other resin suppliers being aware of this new specification may wish to compete and, if so, will offer their version "second-time around".

If, for example, three nylon suppliers have products that meet with approved Ford specifications, the molder will probably decide which one to choose. Ford really does not care, and, that is one reason why they prefer several competing products - more for the molder to choose from.

Often, an automotive firm will tell a resin supplier, "we are not interested, at this time, in a new product, the current one is working out very well" - a frustrating reply to the plastics supplier.

The ASTM also has "line call-out" numbers associated with physical, chemical and electrical properties of a specific plastic - irrespective of its application or automotive company specification. Therefore, each automotive plastic grade has an automotive company and ASTM number.

As an example of the complexity of this scenario, the Allied-Signal Engineering Plastic Automotive Specification Guide - 1993 lists about 40 ASTM 04066 "line call-outs" for its CAPRON and NYPEL nylon products. Chrysler Corporation has over 50 and Ford over 20 etc.

In order to reinforce the point regarding technical cooperation between plastics suppliers and the automotive industry, a review of a very current trade journal article is appropriate. This is a particularly important event since it is a long-term arrangement.

Ford Motor Company and General Electric are joining forces in a 5-year/$11 million venture to demonstrate the ability to produce structural composites from cyclic thermoplastic polymers.

The National Institute of Standards & Technology (formerly the Bureau of Standards) is providing about half of the funding.

The work will be done by Ford's Research Laboratory in Dearborn, MI and GE Research & Development Center in Schenectady, N.Y. Several Ford and GE contractors will also be involved in the project.

This joint-venture will try to develop a generic automotive component (e.g., structural cross-member) under automotive manufacturing conditions.

The federal government would like to commercialize advanced technologies as a way of maintaining a U.S. competitive position.

VIII - COMPETITIVE ANALYSIS

A - Plastics Producers

The decline in plastics usage corresponding with weak auto demand has not picked up appreciably by the end of the 2nd quarter of 1993. Many plastics producers feel that the automotive plastics markets will rebound, albeit slowly. As a result, suppliers in a slow-moving economy will increasingly prefer to spread the risk mostly via joint ventures when considering expanding capacities or spending money on new technologies.

This tactic will accomplish several objectives: pooling of resources both financial and technical, increasing global breadth, and operating at a reduced cost base. These foreseen arrangements will probably extend to collaboration on manufacturing and/or production whereas one company would build a monomer plant, the other a polymer plant.

Simple forms of these arrangements have already occurred by swapping which allows both to broaden their respective product lines and test market prior to building a plant.

Table 40 summarizes the areas of concentration of major automotive plastics suppliers.

TABLE 40

SELECTED AUTOMOTIVE SUPPLIER POLYMER AREAS OF CONCENTRATION

---------- Polymer Type ------------

Supplier	A	B	C	D	E	F	G	H	I	J	K	L	M	N	O	P	Q	R	S
Borden			X																
Allied-Signal						X					X						X		
Chevron	X																		
BASF		X	X	X			X	X			X		X	X					
Dow	X		X								X	X					X	X	X
Dupont						X					X		X					X	X
Eastman	X	X		X													X		
GE Plastics				X		X					X	X		X	X	X	X		
Hoechst-Celanese	X	X				X					X				X	X	X	X	
ICI									X										
Miles									X		X	X			X		X		
Monsanto				X							X		X		X		X		
Phillips	X	X														X			
Elf Atochem			X				X				X						X		
Himont		X																	
Arco		X	X																
Aristech							X		X										
Ashland							X	X	X										

Code:

A = Polyethylene
B = Polypropylene
C = PVC
D = Polystyrene/Styrenics
E = Other Commodity Polymers
F = PET/PBT
G = Unsaturated Polyesters
H = Phenolics/Epoxies
I = Polyurethanes
J = Other Thermosets

K = Nylons
L = Polycarbonates
M = ABS
N = Polyacetals
O = PPS
P = Other Eng Polymers
Q = Polymer Alloys/Blends
R = TPEs
S = Other

Source: TPC Business Research Group

B - Concept of Product Substitution

Although this study has indirectly touched upon the subject of
product substitution (or competition between automotive
materials), this section will review this topic, add additional
information and provide future possible scenarios.

Powder metal parts are reaching the point where they can compete
with nickel/molybdenum steel alloys. Will this new development
have a significant effect on possible plastic substitution for
these steels? Applications for these new powder metal materials
include: synchronizer hubs, blocking rings, transmission chain
sprockets, parking gears, connecting rods, etc.

Instead of being cast, forged and machined, powder metal
components are formed by pouring fine metal powder into molds,
compressed at very high pressure and then hardened in ovens.

The Big Three automakers are switching cast iron to aluminum
engine blocks, along with Honda and Subaru. The SATURN, INFINITI
and LEXUS models have engines with aluminum blocks.

However, aluminum intake manifolds appear destined to be replaced
by thermosets and thermoplastics.

Over on the "steel side" the trend seems toward building larger
vehicles - somewhat "against the grain" in terms of higher CAFE
requirements. If this is the case with some models bigger cars
mean more steel, and include the following recent models:

--- Cadillac EL DORADO
--- Ford TAURUS
--- GMC SUBURBAN
--- Buick SKYLARK and LESABRE

139

These models are longer and wider than previous versions and contain deeper front side rails and crossmembers for added strength and stiffness.

An example of the extent of intra-material product substitution in the automotive industry is the Dodge VIPER two-door sports car. Although the VIPER has limited first year production to about 5,000 units, its multi-material make-up may be a portent of the future as follows:

--- 160-200 lbs. of plastics for body and structural applications
--- 300 lbs. of aluminum - (four to five times as much as other U.S. cars)
--- 6 lbs. of magnesium components
--- stainless steel exhaust manifolds
--- hot-dipped galvannealed steel for its body frame

Intra-plastic competition in automotive applications is another important production substitution phenomenon.

The following trends have a direct bearing on the types of polymers that will be used in automobiles and trucks:

--- new designs and shorter design cycles
--- recyclability
--- parts consolidation
--- environmental/safety concerns
--- type of fuels
--- more pronounced cost pressures
--- changes in technologies

The polymer decision varies with each specific application. The following lists each major automotive application with corresponding polymer choices, although most competition is between steel and reinforced plastics.

Exterior Body Panels: urethane, polyurea, thermoplastic polyesters, nylons, ABS, alloys/blends (PC/ABS, PPO/nylon). There are a multitude of process options available which include: SMC, SRIM (PU, polyesters), RRIM, RIM, RTM, injection molded resins (reinforced and unreinforced), composites (vinyl esters, polyesters), TPEs.

Critical factors affecting a decision can include: Class A Finish, physical properties, ability to paint "in-line", ability to withstand collisions, etc.

Competition was very intense for GMC's SATURN in which the following were considered: injection molded alloys, PU and polyurea RIM, SMC (polyester), RTM (PU and polyester).

The final selections were:
--- doors (PC/ABS)
--- fascias (TPOs)
--- front fenders and rear quarter panels (PPO alloys)

Bumper Systems:

--- bumper beams - alloys/blends, PC/PBT, polypropylene, vinyl
    esters, PUs, SRIM, RTM, polyesters
--- bumper covers/fascia - alloys/blends, PC/PBT, polypropylene,
    PU, polyureas, RIM
--- energy absorbers - PU foam, polypropylene, ethylene vinyl
    acetate

Competitive factors in this automotive application include: impact resistance, in-mold color ability, design flexibility, lowered weight, etc.

Interiors:

--- PUs, polyesters, alloys/blends, polycarbonates, ABS, polypropylene

Further breakdown into smaller segments reveals the following competing polymers:

--- IPs - polypropylene, PC/ABS, ABS, polycarbonates, SMA
--- seat padding - PU foams, PET fibers
--- consoles, door trim & others - ABS and polypropylene

Factors affecting polymer choices are: part processability, mold colorability, acoustics, aesthetics, etc.

Another way of looking at the possibility of product substitution is to examine the plastic automotive applications in more specific detail. The problem with this technique is that most plastic suppliers list an almost limitless list of "possible" applications. One example will suffice: Miles, Inc., in their trade literature has a two-page spread showing an auto with specific areas being used/suitable for their products (Table 41).

C. Molders

The business scenario surrounding automotive molders is focused on two major factors - cost cutting by the automakers and free trade.

Several recent developments have appeared during 1992 that will directly affect automotive molders. The first being the previously discussed global purchasing consolidation of General

Motors. Basically, GMC is "market testing" all of its contracts to re-establish standard pricing which will be significantly lower than current levels.

The second factor is the North American Free Trade Agreement (NAFTA) which will result in a doubling of auto production in Mexico within three years following ratification. Should molders consider joint ventures with Mexican companies as a beginning strategy for market penetration?

Thirdly, the probability of increased domestic sourcing by Japanese transplants due to narrowing of culture gap between these automakers and North American molders. Molders must learn, however, to allow for lengthy approval processes and permit careful scrutiny from transplant managers.

In regard to the General Motors scenario, molders complain that GMC is calling for over a dozen quotes per contract, soliciting offers from less experienced molders. Both Ford and Chrysler, will not follow GMC's lead, and plans to develop long-term "partnerships" with a pared down list of suppliers. It is very unlikely that either Ford or Chrysler would open up existing contracts to bidders.

One fallout from General Motors' recent action is that the ability of a supplier to finance long-range advanced programs has changed very dramatically. Molders have little assurance, when participating in negotiations, that they will have an opportunity to recover their investment. One comment was that molders may have to decide whether they want to become a leader in technology or a low cost resin molder.

In regard to NAFTA, molders need clarification on several of its provisions especially the complicated system for tracing produc-tion sites of parts made by the various layers of subcontractors. What happens when a supplier produces a subassembly from imported

parts? Will only the value of the assembly work be counted as North American?

Currently, large captive plants are being set up by the automakers in Mexico, in addition to opening of facilities by large automotive subassembly producers. Realistically, only relatively large molders will go to Mexico to establish operations.

Molders who do begin operations in Mexico will opt for Mexican partners since the law requires minimum Mexican share in owner-ship of any company based in that country.

As to improved relations with Japanese transplants, increased joint-ventures between Japanese and North American companies have paved the way. Increased cooperation will also result from transferral of parts engineering from Japan to North America. Access to all parts of the molder's operation by Japanese transplant personnel is critical to establishing long-standing relationships.

Another technique for molders to have Japanese transplants increasing their domestic sourcing is to enter into joint-ventures. This strategy, however, has met with little success, so far.

In the final analysis, the number of molders serving the automotive industry will continue to decrease because of the three issues just discussed. Molders will need sufficient "critical mass" to deal with General Motors' actions, NAFTA and the increased relationship with Japanese transplants.

From the automaker's standpoint, a reduced number of primary suppliers is a key objective. Chrysler had about 3000 suppliers years ago, and is down to about 2500 - the goal being less than 1000 - by the mid 1990s.

The automaker's rationale for a smaller number of larger suppliers rests on three points - improved product quality, better service and lower costs. The Big Three feel that they will gain these objectives from large, talented and aggressive companies.

These larger suppliers should be more capable of investing in R&D, taking on more responsibility for designing, developing, manufacturing and assembling large components to auto assembly plants.

IX. Technology

A. Polymer Fabrication

There are two broad classes of thermoplastic resins - amorphous and crystalline. The latter are characterized by melting and freezing points, the former are not. Examples of crystalline thermoplastics include HDPE, LDPE, polypropylene, polyacetals, nylons and polyesters. ABS, cellulose acetate, polyphenylene oxide-based resins, polycarbonates, PVC, polystyrene, SAN and some polyethylene are important examples of amorphous thermo-plastics.

About half of all thermoplastic products in use today are prepared via extrusion. In this process, the polymer is melted in an extruder and forced through a die which, in turn, converts the melted material into the desired shape. When the melt is cooled, the shape is maintained.

When the molten thermoplastic is injected, under high pressure, into a steel mold - an injection molded product results. When the plastic solidifies, the part takes the shape of the mold cavity. Another major injection molded process is blow molding, which is the most common process used to make hollow parts. Thermoforming is often used to convert plastic sheet into parts, and, if under a vacuum, the process is known as vacuum-forming.

Acrylics are usually sold as cast sheeting, and polystyrene is often available as expandable molding for foamed products. Foamed polystyrene sheet extrusion is another commercially important product.

Calenders have heavy steel rolls, and this type of equipment is used to make sheeting from semi-rigid and flexible thermoplastics such as PVC and ABS.

147

Many thermosetting polymers are processed by compression molding
which is the oldest plastic processing technique. A compressed
"cake" of the granular resin is placed in a mold, at which time,
the resin melts and the pressure forces the molten material to
fill the mold cavity. The resin is cured by continued heating
and the final product is removed from the mold.

Most unsaturated polyesters are reinforced with glass fibers.
Polyurethanes are often made via reaction-injection molding (RIM)
to produce partially foamed materials. This process consists of
the rapid injection of liquid streams of polyols and an
isocyanate into a heated mold.

One area of plastics technology in which interest has grown
dramatically is liquid composite molding (RTM) and its speedier
counterpart SRIM.

The automotive industry's emphasis on higher productivity and
lower-volume market niches has spurred interest in these methods
of relatively low-cost production of fully consolidated composite
parts.

RTM and SRIM low tooling costs and "design freedom" may provide
automakers with a good fit with their preoccupation with
quick-model changes.

Closed-mold RTM and SRIM would be more environmentally friendly
than polyester spray up processes.

With part molding cycles at about 30 minutes, RTM is currently
restricted to low-volume market, e.g., heavy trucks. Heavy truck
cab exteriors are prime candidates for RTM.

SRIM is expected to move from smaller, non-appearance parts into front end structures (composites) such as bumper systems.

B. Plastics Product Performance/Development

As previously mentioned, there are specific performance requirements that must be met by automotive plastics. For example, thermoplastic exterior parts must withstand on-line paint temperatures, have relatively low thermal shrinkage and yield Class A surfaces. The heat resistance of these thermoplastics must be sufficient for them to pass through the highest paint temperatures.

Lower bake temperatures would enable thermoplastics to possibly replace SMC and RIM materials in addition to replacing metals.

Generally, all automotive plastics must have high-impact strength, chemical resistance and durability.

Other performance requirements for automotive plastics are materials development and handling. Manufacturing large, complex plastic parts at sufficient rates to meet a 50+ car/hour production line has, and will push process technology and equipment to the limits of present capabilities, although compression molding can handle the sizes needed for structural parts. Extensive new equipment is needed for major chassis molding in the 5,000 ton class. RIM is effective with large parts, but must have relatively short cycle times, while structural RIM (SRIM) and high speed transfer molding processes are under investigation and hold some promise.

"Ordinary" type of injection molding has produced Class A finishes on moderate-sized panels, but a major objective is to have this process effective for high speed production of large plastic parts.

149

Within this decade, use of plastics and composites will pass
through several design/engineering phases:

--- fixed and movable skin panels, requiring a Class A finish to
    be used on load-carrying frames.
--- manufacture of fixed horizontal composite-based surfaces with
    requisite stiffness and be able to withstand torsion and
    bending loads.
--- availability of structural composites for the rear suspen-
    sion, trunk floor, body pan, etc.

There are still several critical issues that need to be resolved
before plastics can move into the "next phase". Some of these
include:

--- standardized tests and specifications to ensure high
    performance in production and service.
--- need of tools which can fabricate large, complex components
    with reduced forming times.
--- emergence of systems to recycle (or dispose) of automotive
    plastics after they are disposed of.

Standards and test methods define the properties of plastics, and
are a critical factor in their applications. This is especially
true when additives are incorporated into polymer formulations
because they change the physical and chemical characteristics of
the original substrate.

Therefore, properties of formulated polymer products are
determined by their chemical composition, molecular weight,
structure and morphology.   The results of testing are critical
for quality control, design and R&D.

Mechanical tests include, tensile strength, impact-strength, elongation, tear strength, abrasion resistance, etc. Most tests are generally grouped into four categories: general, mechanical, thermal and electrical.

There are several sources/agencies that conduct tests on plastic materials such as:

--- The ISO Technical Committee on Plastics has formulated over a hundred standards which basically describe analysis and evaluation methods for plastic materials.
--- The Department of Commerce publishes The World Index of Plastics Standards which contains almost 10,000 national and international standards on plastics and related materials.
--- The Defense Department has an Index of Specification and Standards (DODISS) which lists unclassified federal, military, and other specifications and standards. The section entitled (Group 9330) deals with fabricated plastic materials which includes resins.

There are a myriad of technical advances being made some of which are touted by plastics suppliers, molders and the automotive industry. No one report could possibly cover all of these cited in the trade press and interviews. The following is a selected representative list by general application areas that were reported in the last twelve months.

Interior:

--- auto floor sections and interior panels covered with carpeting can now be made in one-step via injection-molding which could have an effect on materials chosen.
--- plastic/metal composites applications for loadbearing auto components such as IPs and doors.

151

--- development of stain-resistant, semi-flexible polyurethane
   foam systems for vinyl-covered IPs.

--- slush molding becomes an option for interior padded parts.
   Automakers searching for dashboards that can be made out of
   single-family materials to replace multi-materials to ease
   recycling. This process used to make PVC IP skins may yield
   first commercial thermoplastic polyurethane skins for padded
   dashboards.

Exterior Body:

--- compression molding advances improved SMC and BMC processes
   and those of reinforced thermoplastics have been extended to
   cover integrated assembly operations such as rear doors.

--- truckhoods with Class A surface from thermoset SMC via single
   directed fiber perform a new low-profile polyester resin
   system and new variation of liquid composite molding-cures
   in 2 minutes putting in range RIM systems for large parts as
   well as compression molded SMC. Large resin transfer
   molding (RTM) parts usually have 20-30 minutes molding
   cycles.

--- cold-gas polishing, a technique using liquid nitrogen and a
   polishing compound fixes flawed paint on plastics. Process
   claimed to be superior to polishing and buffing because
   friction-induced heat can irreversibly damage highly flexible paints used on plastics.

Other Exterior:

--- Ford is testing polycarbonates and acrylics from GE Plastics
   and Rohm & Haas, respectively to evaluate replacing glass
   with coated plastics and probably limited to rear windows
   and side glass and sunroof panels.

--- one-side RIM encapsulation, a new technology for automotive
   window design has been installed in the Ford ECONOLINE Club
   wagons and vans.

Fenders/Bumpers/Fascias:

--- development of a continuous system for plasma-treating
   polypropylene bumpers on a 50-second injection molding cycle
   at lower cost than flaming or UV treatment.
--- structural composites helping to reduce bumper weights on
   Buick ROADMASTERS
--- bumper beams made of SRIM able to meet 5 mph impact standards
   without added material and weight of competing composite
   technologies.
--- one-piece beam made from polycarbonate replaced two-piece box
   section made from glass-reinforced polypropylene.

Under-the-Hood:

Custom sheet processors are beginning to make an impact on the
automotive market. Wider ranges of thermoformable materials,
including high performance grades allow custom sheet processes to
challenge both molding as well as extrusion processes. The five
largest custom sheet processors in North America are listed in
Table 41.

High-molecular weight polyethylene has been the most popular
automotive material for sheet processors, primarily for truck bed
liners and fender shields. Other polymers are being used and/or
considered for dashboards, interior panels and spoilers comprised
of ABS, PVC and polypropylene. Engineering polymers are being
investigated for possible future structural applications, e.g.,

TABLE 41

THE FIVE LARGEST CUSTOM SHEET PROCESSORS IN NORTH AMERICA

Company	Approximate Capacity Million lb/yr	Remarks
Primex Corp	185	Processes primarily polystyrene in six locations for a wide range of applications.
Spartek Corp	185	Emphasizes ABS, but processes a variety of materials at six plants for various markets.
Pawnee Extrusion Systems	100	Emphasizes ABS, but processes a range of sheet for various markets.
Panda Corp.	100	Processes primarily HMW-HDPE sheet for automotive applications.
United Technologies Automotive	60	Processes primarily HMW-HDPE sheet, strictly for automotive applications.

Note: This list does not include sheet processors directly linked to materials supply companies, or those dedicated to finished product lines.

Source: Modern Plastics

154

polyetheretherketones, and liquid crystal polymers. One example being body panels and bolt-on bumpers from polycarbonate/polyester blends, and polycarbonates.

A trend from metal stamping to plastics is envisioned by many in the sheet processing business for "hidden" parts that shield more sensitive metal parts from the environment.

Interest in resin transfer molding (RTM) has grown, fostered by the auto industries emphasis on higher productivity and lower market niches. Increased design freedom and lower tooling costs fit in with the auto industry's quick-model change philosophy. Increased concern about VOCs has also spurred interest for closed-mold RTM.

RTM is currently restricted to lower volume markets because of its 30-minute part-molding cycles. A major outlet being heavy truck cab exteriors.

The RTM processes need to be speeded up. In this are placed relatively slow process, low-viscosity polyesters or epoxies into a mold containing a reinforcing preform (usually glass) via static mixing. At least ten minutes are needed to dispense the resin and catalyst to the mold.

SRIM, a competitive liquid composite molding process, was covered in the section on Polyurethanes. However, as liquid composite molding materials expand, the distinction between RTM and SRIM begins to blur.

Arc discharge lighting, a polymer-based lighting system is expected to become a standard in cars by the end of the decade. It will replace conventional bulbs with a central high-intensity lighting source and fiber-optic transmission.

Two acrylic rods containing both high and low beams in a one-inch by six-inch package will replace traditional car headlights, e.g., a line-of-light. This system generates much less heat than headlamps which is a positive factor for plastics.

The light is actually drawn through plastic (acrylic or polycarbonate) or other polymeric optical fibers from the source - a high-intensity arc discharge light which employs a continuous electric arc between two electrodes in vaporized metal. GMC is considering installation of these arc discharge lights on some of its 1995 models.

A plastic optical fiber, INFOLITE, made by co-extruding a core of polymethyl methacrylate and fluoropolymer cladding has been developed by Hoechst-Celanese.

Polymer fiber optics are less cumbersome than shielded copper cable and would also reduce wiring weight by over 50%. Glass fiber optics are more costly, sensitive to vibration and breakable according to Hoechst-Celanese. The product has potential in the automotive market but is not suited for under-the-hood applications. INFOLITE is not as efficient as glass fiber in regard to light transmission, but can be used for the relatively short lengths for auto applications.

Traditional blow molding has expanded its role in the automotive market based on lower tooling costs, inherent parts consolidation and integration of secondary operations. A polyethylene windshield washer fluid reservoir and a high-heat UV-stabilized polypropylene pressurized radiator supply bottle are just two examples.

Single blow-molded units that perform multiple functions while consolidating separate parts is a growing trend with auto companies. One important example is a new instrument cluster hood on a recent Pontiac GRAND AM.

X. Environmental/Regulatory Considerations

A. Overview

This enormous and far-reaching topic has been in the trade press for decades. This section of the report will briefly review important past events, assess the current climate and provide several possible scenarios for the next five years.

It should be noted that automotive engineers rank the lowering of greenhouse emissions, reduced hydrocarbon tailpipe emissions and improving fuel economy as the major environmental priorities. Most plastics suppliers view fuel economy as the number one issue, followed by hydrocarbon emissions and reducing pollution from manufacturing operations.

Both groups placed toxic waste and vehicle recyclability well back on their priority lists.

Several within the industry suggest that there is too much politics and not enough technology with current efforts. "Legislation seems politically motivated by public response and not by those who know how to direct automakers how to profitably invest in environmental change".

Hydrocarbon-emission control, fuel economy and plastics recycling are three issues which will obviously impact the automotive industry. This section will focus on the latter since the first two were covered in an earlier section of this report.

Disposal or retrieval of plastic components in junked cars is a concern of automotive plastic suppliers. Several companies have already established repurchasing and recycling facilities. A

157

somewhat unrealistic objective of the entire plastics recycling program being that plastic components should be clearly marked as to type of polymer - easily readable to the public.

In addition to the EPA which oversees the plastics recycling scene, the National Highway Traffic Safety Administration (NHTSA) set into motion new rules which will require that cars sold in the U.S. conform to new side-impact crash tests. Front end tests have been in place for several years. These new standards will be phased-in over a four-year period beginning September 1st, 1993.

Until now, there was no side-impact crash test, only "static" tests requiring car doors to resist force applied by a piston pressing a steel cylinder against them.

In the new test, a moving barrier simulates a vehicle striking a test car from the side. Since this test is dynamic, dummies will be used to measure potential injuries to passengers.

The automotive companies are free to meet these impact standards any way they feel best fits the design of the car. There was some controversy over the NHTSA's side-impact dummy (SID), and GM is trying to get its own model accepted.

While some vehicles already meet the new side-impact standards, others will need to add foam padding, door beams, or other means to absorb the energy. Design changes will probably be in the offing. A most intriguing one is side-restraint air bags - either coming from the door panel or the center rest arm area.

Side-impact crashes often cause internal injuries (thorax and pelvis) and the next step in this process will be geared toward protection of the head. Early suggestions include laminated glass and softer interiors.

Another emerging environmental issue relates to adequate ventilation systems and more acceptable SMC disposal methods facing molders. More effective styrene fume handling systems are anticipated.

Clearly, diminishing landfill capacity makes it imperative that scrap SMC be disposed of in another manner. Pyrolysis of SMC yields methane and ash, the latter may have potential as filler material. Another possibility is grinding SMC for use as boiler fuel.

B. Recycling

Soaring landfill costs and increased environmental awareness have turned car recyclability into one of the largest engineering issues of the decade, the major participants being material suppliers and automakers. This issue has already influenced design and material choices.

A recent Society of Automotive Engineers (SAE) annual show unveiled all-aluminum cars primarily because of the metal's advanced recycling profile compared to plastic. Several consider plastics as a primary roadblock in making cars more recyclable, but it is undergoing changes to make them more environmentally-friendly.

It is estimated that 75% of a vehicle's weight is currently recycled, but it is almost all metals, e.g., steel, cast iron, and aluminum parts which are easily identified and salvaged.

However, that leaves hundreds of pounds of other materials such as glass, fabrics, and plastics that go largely unclaimed. It is considered almost impossible to salvage windshields (plastic/-glass laminates) and rear windows. The biggest problem is still the 200+ pounds of automotive plastics on an average domestic car.

The major automotive plastics recycling success story is the recovery of polypropylene from auto battery cases. A major domestic polypropylene supplier claims that over 95% of all U.S. batteries are recovered for their polypropylene and lead components. About 40% of reclaimed polypropylene goes into new automotive batteries, the remainder is used in other automotive and consumer products.

About a hundred different types of plastics are used on cars and light trucks, and very little of it is recycled because it is very difficult to identify and sort the different plastics. Attempts to mix and melt dissimilar plastics into useful new products have been largely unsuccessful.

Considering that over ten million cars are scrapped each year, "waste" plastic volumes are estimated at about one million tons per year - assuming 200 pounds of plastic per car. Actually, this volume will not be reached until the non (or low volume) - plastic cars have been scrapped. Other sources place the volume of ASR at 2.5-3.0 million tons.

Most plastics recycling centers are with the packaging segment - mostly PET soda bottles and HDPE bottles (including milk jugs). Estimated percentage of post-consumer recycled plastics reached about 3% in 1992.

When compared to the total solid waste problem, one million tons is a very small factor. In fact, total plastic waste comprises only 10% of the total solid waste volume.

What aggravates the automotive plastics recycling problem in the public's eye is the size and visibility of the autos themselves.

Several skeptics have said that keeping this issue in the front pages is wonderfully cheap publicity for the automotive industry,

along with their plastic suppliers.

The strongest automotive recycling movement is in Germany, which has mandated that all materials used in car production be recycled by the year 2000. The entire European Economic Community may also mandate that automakers be responsible for recycling their own cars. Germany has a national network of authorized independent vehicle dismantlers that return sorted regranulated scrap plastic to car manufacturers or material suppliers.

Currently, there is no legislation in the U.S. that requires recycling of plastics, or any other material from automobiles.

Steel has been recycled for years, since there is a well-established infrastructure for collecting and claiming steel. Iron and steel auto parts can be easily separated at salvage yards - not so with plastics.

The aluminum industry, in addition to steelmakers, is jumping on the recycling bandwagon in its attack on plastics since aluminum has a far more advanced recycling profile. It is claimed that 60%-80% of aluminum automotive scrap is currently reclaimed and recycled.

Two recent events are noteworthy. The first is the use of recycled SMC in General Motors' Chevrolet CORVETTE for 1993. This will be the first instance of a North American car using recycled SMC body panels. The part is the reinforcement around the rear window that supports the rear portion of the roof. The piece is not visible because it is covered with trim. The grinding technology used was developed from work done by the SMC Automotive Alliance. The recycled material was compounded and molded by GenCorp Automotive.

The second event consists of Toyota and Fiat proposals for car part recycling solutions. The two companies have launched a

pilot program to collect and recycle polypropylene bumpers into trimmings, sill board, heater housings, dashboard air ducts and air filters. After several applications, the recycled plastics having lost most of their properties can be used as fuel in smelting furnaces.

The idea of recycling the number of polymer families used is receiving more and more attention. In some cars, there are over 20 resin types used and many automakers would like the number to be less than ten. This occurrence would certainly simplify recycling efforts.

Polypropylene, which is readily recycled by commercial processes is a leading candidate for "standardization". Suppliers, understandably, tout polypropylene citing that it can be supplied as homopolymer, copolymer or terpolymer resin and is suitable for alloying/blending, can be compounded as an elastomer or produced as foams, and accepts fillers and reinforcements.

There is a drive to eliminate some materials that are used to fabricate small parts (polyacetals, nylons), that may not be compatible with other more predominant resins forming most of the part. These materials hinder recycling and raise disassembly costs since they must be removed before collection.

The pressure to recycle might allow PET to erode the SMC market. A long glass fiber-reinforced PET sheet has already been introduced as an SMC alternative. This material is claimed to be about 10% - 20% lighter than SMC and can be formed on SMC machines.

Automotive plastics suppliers counter by stating that their products' advantages of weight reduction and increased design flexibility outweigh the "current" recycling disadvantages. Both the automotive Big Three and plastics suppliers have joined forces with developmental efforts to improve the recyclability.

162

The SAE has announced a system for identifying plastic parts
(J1344 codes) to help in recycling. The automotive Big Three
plastics suppliers and processors endorsed this action.

This system uses acronyms to differentiate among over 120
different thermoplastics and thermosets. Fillers, alloys,
laminates, etc. add to the complexity of the situation.

A critical issue receiving a great deal of attention is how to
reduce the number of different plastics per car. Such an event
would go a long way to ease the recycling problem, however, this
would decrease demand for virgin resin. Others feel that unless
the plastics recycling program moves ahead, other non-auto
markets may choose different materials, as well.

Furthermore, the Big Three are expected to announce the formation
of a consortium - Vehicle Recycling Partnership (VRP) - to deal
with recycling. More recently, the Automotive Group for the
Partnership For Plastics Progress is moving ahead with major
initiatives to address the environmental issues related to
automotive plastics, especially recycling. The group represents
the leading resin producers and is supported by the SPI.

A major thrust of this group is to develop an infrastructure for
the recovery and recycling of "post-consumer" automotive
plastics.

The plastics industry cites the following advances in this area:

--- thermoplastic parts can be remelted and formed into similar
    parts, e.g., fenders, bumpers.
--- recovery of polypropylene from automotive batteries.

--- RIM thermosets (which cannot be remelted) are ordinarily land-filled from junked cars, can be recycled at a 10% rate, e.g., seat shells. However, cryogenic grinding is still not practical although progress has been made in terms of new approaches such as allowing 10% of some regrind to be added to virgin RIM polyurethane or compression molding of 10% RIM regrind into finished parts. Mechanical properties of regrind not as good as virgin resin but suitable for non-appearance applications. Another approach is to recycle polyurea elastomers via thermal processing granular waste material into extruded automotive trim.

--- programs being initiated to take existing thermoplastics out of the waste stream profitably.

--- research underway to breakdown polymers into its chemical building blocks (monomers), e.g., DuPont's "methanolysis" or breakdown of PET.

--- investigating ways to produce new products from scrapped SMC.

The recycling system is fairly complex and begins with automobile dismantlers who remove reusable parts, radiators, catalytic converters, bumpers and components that must be removed before shredding by law (batteries, fuel tanks).

The remainder of the car is shipped to auto shredder companies who break them down into reclaimable metal pieces or automotive shredder residue (ASR).

Designing future cars for easier disassembly would ease the recapture of plastics provided they were properly coded. In-mold labeling of auto parts is expected to be fully in-force before the end of the decade.

The Polyurethane Recycling and Recovery Council is examining ways to remove polyurethane foam from car seats prior to shredding - a technique which will reduce the volume of ASR.

Viable recycling technologies continue for polyurethanes, but several challenges remain. Currently, rigid foam recycling is developing more slowly than flexible foam recycling.

It is estimated that post-consumer rigid polyurethane foam will reach 250 million lbs by 1995. Scrap will constitute 30 million lbs or 12% of the total. The possible recycling/recovery methods include:

--- compressed parts - using rigid scrap to make new parts
--- filled foam - to be used as possible filler in production of roof and wall insulation
--- flexible bond underlay - in carpets
--- paint sludge removal - using rigid foam dust for detackifying paint

C. CAFE Changes

In an earlier section of the report, the origins and current status of CAFE requirements were discussed. This section reviews possible future scenarios.

CAFE increases by 20% by 1996 and 40% are possible according to several observers, but would require an enormous effort and affect car prices. It is questionable whether car buyers would want the resultant vehicles since they would be much smaller and more expensive. Would the public go for a small 50 mpg car?

A major question which has yet to be resolved, is what percent of the automaker's R & D budgets will be committed to CAFE increases?

In regard to the two-stroke engine, "no one really knows" whether it will have an impact or be a "technical flash in the pan".

XI - COMPANY PROFILES

ADVANCED ELASTOMER SYSTEMS - ST. LOUIS, MO

Advanced Elastomer Systems (AES) is a joint-venture between
Monsanto and Exxon Chemical. AES supplies thermoplastic
elastomers (TPEs) and related products as follows:

--- SANTOPRENE - thermoplastic rubber for highly-engineered
     applications designed to replace polychloroprene rubber
     (PR), and EPDM rubber compounds.
--- GEOLAST - TPEs with enhanced oil resistance that meets or
     exceeds that of nitrile rubbers.
--- TREFSIN - TPEs with superior permeability resistance for
     medical, industrial and consumer products.
--- DYTRON XL - TPEs with high performance material for wire and
     cable market as alternatives to CR, EPDM and cross-linked
     polyethylene.
--- VYRAM - TPEs with mid-range performance to replace EPDM,
     SBR and natural rubber.
--- VISTAFLEX - TPEs meets demanding appearance requirements to
     replace styrene-block TPEs and plasticized PVC.
--- TPR - thermoplastic rubber for economical part performance.

Advanced Elastomer Systems has R&D facilities at Akron, OH and
Louvain-La-Nueve (Belgium), and manufacturing operations in
Akron, (OH), Pensacola (FL), Brazil, Japan and Wales.

Selected SANTOPRENE automotive applications include: weather
stripping, electrical wire and cable covering, hoses, tubing, air
ducts, gaskets, sleeves, grommets, steering system boots, etc.

AES literature recommends that the automotive industry can evaluate SANTOPRENE for the following: engine, electrical system, passenger compartment, front wheel drive axis/steering, fuel system, brake systems, AC/heater, automotive transmissions, doors, air intake and emission systems, front suspension, bumpers, cooling system, power steering unit, wheels, rear axle and suspension, vacuum pump, and windshield wiper/washer systems.

Company literature also lists light truck and van applications for VISTAFLEX, VYRAM, DYTRON XL and GEOLAST.

ALLIED-SIGNAL, INC. - MORRISTOWN, NJ

Allied-Signal provides several plastics to the automotive industry which include:

--- CAPRON (nylon) - general purpose, plasticized, flexible and impact resistant, mineral reinforced and via extrusion/blow and rotational molding for the following selected applications: gears, connectors, tubing, jacketing, cooling fans, shrouds, vacuum reservoirs, window hardware, interior light components, power steering reservoirs, etc.

--- NYPEL (regenerated nylon) - general purpose and mineral reinforced bushings, seat belts components, dome lamp bezels, fans, timing belt covers, etc.

--- PETRA (PET) - reinforced grades for ignition and carburetor components, electrical connectors, switches, vacuum pumps, fuel impellers, air pump housings, external mirror components, etc.

--- DIMENSION - nylon alloy for body panels and fenders, wheel covers, mirror housings, etc.

AMOCO PERFORMANCE PRODUCTS - ATLANTA, GA

Amoco Performance Products is a part of Amoco Chemical which is a division of Amoco Corporation. The company's engineering plastics are important products and include the following:

--- AMODEL - polyphthalimide (PPA) thermoplastic resins available as unreinforced, glass/mineral reinforced and impact-modified.
--- UDEL   - polysulfones (dashboard climate control panels).
--- MINDEL - resins (decorative parts).
--- RADEL - polyarylsulfones (little automotive applications).
--- ARDEL - polyarylates (high-level tail lights).
--- TORLON - polyamide-imide (valve stems).
--- KADEL - polyetheretherketones (PEEK) (little automotive application).
--- XYDAR - liquid crystal polymers (electrical systems).

ARCO CHEMICAL CO. - NEWTOWN SQUARE, PA

Arco Chemical, a subsidiary of Arco, Inc. (formerly Atlantic Richfield) supplies several polymers to the automotive industry which include:

--- ARPRO - polypropylene foams and ARPAK polyethylene foams via ARCO/JSP - a joint venture of Arco Chemical and Japanese Styrene Paper (Mitsubishi Gas Chemical America). These products are used as bumper energy absorbers.
--- DYLARK - engineering resins in transparent, impact and glass-reinforced grades with applications in: instrument panels, consoles, headliners, glove box doors, and trim parts.

--- ARLOY - blends of DYLARK styrene copolymers and polycarbonates used in one-piece instrument panels, seat belt
  retractor, housings and speaker grilles.
--- DYTHERM - expandable copolymers with several automotive
  applications, e.g., large temperature resistant, energy
  absorbing shapes with a variety of skinning options, such as
  door liners, IP components, wheel covers, consoles, sun
  visors, arm and head rests, heater ducts, etc.
--- DYLITE - expandable polystyrene recommended for side door
  panels.

ARISTECH CHEMICAL - PITTSBURGH, PA

Aristech Chemical, a Mitsubishi subsidiary, provides polyester
resin systems for SMCs, TMCs and BMCs. The company's products
are all tradenamed MR followed by a series of numbers, e.g., MR
13006, MR 13017, etc. These materials include isophthalics and
vinyl toluenes under a variety of grades such as reactive,
weather/corrosion resistant, resilients, etc.

ASHLAND CHEMICAL CO. - COLUMBUS, OH

Ashland Chemical, an Ashland Oil subsidiary, is an important
plastics supplier to the automotive industry.

Beginning with the company's development of cold box binder
technology which had an enormous impact on the foundry industry,
Ashland has become a leader in SMC body panel resins and
technology. The company also introduced the technology for
structural composite production by reaction injection molding
(SRIM).

Ashland has been a leader in creating new and approved adhesives and sealants for bonding automotive composites.

Some of their current products include PLIOGRIP structural adhesives, GLASGRIP glass sealant systems, VIC coatings, etc.

Actual plastics or resins supplied by Ashland include the following:
--- AROPOL PHASE ALPHA SMC resin systems - reinforced composite body panels for auto and truck.
--- AROTRAN RTM resin systems - lower volume production of reinforced composite body panels.
--- ARIMAX structural RIM resin systems - for reinforced structural composites.
--- AROTECH structural resins - similar to ARIMAX, but compression molded.

AUSIMONT USA - MORRISTOWN, NJ

Ausimont USA, a Montedison (Italy) subsidiary, is part of Montedison's operating company - Montefluous which produces its fluoropolymer resins (ALGOFLON) and TECHNOFLON fluoroelastomers in Italy.

Ausimont's polymer/elastomer product line consists of the following:

--- ALGOFLON - polytetrafluoroethylene
--- HYFLON - perfluoroalkyl-tetrafluoroethylene copolymer
--- HALAR - ethylene-chlorotrifluoroethylene copolymer
--- HYLAR - polyvinylidene fluoride
--- TECHNOFLON - fluoroelastomer

Montecatini heads Montedison's chemical sector and also owns Himont as well as Ausimont. The Montedison group covers the following business sectors: sugar, edible oils/proteins/animal feed, starch and derivatives, chemicals (22% of the total), pharmaceuticals, and energy. Montecatini's chemical sector includes: polymeric materials, processed and downstream products, advanced materials and specialty chemicals (Ausimont).

AZDEL, INC - SOUTHFIELD, MI

Azdel, a joint-venture between GE Plastics and PPG Industries, manufactures technopolymer structures - a group of fiber-reinforced thermoplastic composites. There are three types of technopolymer structures:

--- AZDEL - polypropylene-based
--- AZMET - composites with crystalline-based polymer matrices
--- AZLOY - composites with matrices of amorphous polymer or
    semi-crystalline alloys and blends

These products are available as continuous random and direction-alized glass mat, chopped fibers and in various colors. Forming parts from Technopolymer structures is accomplished via a compression molding process. Class A surfaces with AZLOY composites via use of high-pressure injection in-mold coating.

BASF INC. - PARSIPPANY, NJ

BASF Corporation, the U.S. subsidiary of BASF AG (Germany), is an important automotive plastics supplier. The company's engineering polymer headquarters is in Parsippany, NJ while its Automotive Plastics Group is located in Troy, MI.

BASF Automotive Polymer business includes four business groups: urethanes, plastic foams, plastic materials and structural materials.

These products have also been listed by the following classifications:

... polyurethanes/elastomers and foams
---ELASTOFLEX - soft polyurethane foams
---ELASTOLAN - TPUs
---ELASTOLIT - structural RIM (SRIM) polyurethanes
---ELASTOPOR - hard polyurethane foams
---ELASTOFOAM - soft integral polyurethanes

... polyurethane raw materials
---LUPRANATE - TDI, MDI
---PLURACOL - polyols

...foamed polyolefins
---NEOPOLEN - polypropylene foams

... plastic materials
---ULTRAFOR - polyacetals
---ULTRAMID - nylons
---ULTRADUR - PBT
---ULTRABLEND - PBT blends
---TERLURAN - ABS
---LURAN - SA
---TERBLEND - ASA/polycarbonate blends
---ULTRANYL - PPE/nylon blends
---ULTRAPEK - polyarylether ketone (PAEK)
---ULTRASON - polysulfones and polyethersulfones
---ELASTOCELL - thermoplastic composites

BASF product literature provides over 50 specific applications for each of their automotive products grouped under interior, exterior and under-the-hood.

BORDEN CHEMICALS & PLASTICS - GEISMAR, LA

Borden Chemicals & Plastics (BCP) is a limited partnership which began operations in November, 1987 after acquiring the PVC resin and basic chemicals business from Borden, Inc. BCP Management, Inc., a wholly owned subsidiary of Borden, owns a minority interest as the sole general partner in the operating partnership.

The Geismar location consists of nine fully integrated plants including a PVC resin plant, along with plants that produce vinyl chloride monomer, acetylene, methanol, formaldehyde, ammonia, urea, and urea-formaldehyde concentrate.

BP CHEMICAL AMERICA - ST. CHARLES - IL

BP Chemicals Advanced Materials Division at St. Charles, IL is part of BP Chemical America (Cleveland, OH) which, in turn, is the U.S. subsidiary of BP, Ltd. (UK) (British Petroleum Company).

The Commercial Composites Group, within the Advanced Materials Division is a major molder of fiber reinforced polyesters and other composite materials and supplies SMCs to the automotive industry.

THE BUDD COMPANY - TROY, MI

The Budd Company, subsidiary of Thyssen AG (Germany) is a leading supplier to the automotive industry. Budd's range of products include steel stampings and assemblies, full frames and chassis frame components, and SMC autobody panels. The latter products come under the aegis of its Plastics Division.

The company recently announced a new SMC (HI-FLEX) which, according to Budd, "bridges the gap between conventional SMC and friendly-type plastics". Budd System 59 was the first SMC plastic molding process to break the one-minute cycle barrier, followed by its TRON-BONDING product which reduces SMC bonding and curing time to about 50 seconds while improving the Class A surface quality of SMC panels.

Budd also provides other plastic automotive components such as doors, decklids, hoods and tailgates as well as engine components such as oil pans and valve covers. Automotive SMC products are made at Carey, N. Baltimore, and Van Wert, OH; and Kendallville, IN.

CHEVRON CHEMICAL - HOUSTON, TX

Chevron Chemical, a Chevron, Inc. subsidiary, produces polyethylene materials with automotive applications for molded products. Specifically, the company supplies:

--- LDPE copolymers - EVA grades PE1220 and PE5285
--- HDPE copolymers - HID 9006, 9012 and 9018

DOW CHEMICAL - MIDLAND, MI

Dow Chemical's Plastics Group is one of the largest suppliers to the automotive industry. The company's major plastics having automotive applications include:

--- LDPE
--- PELLETHANE - TPUs
--- TYRIN - chlorinated polyethylene (CPE) elastomers
--- CALIBRE - polycarbonates
--- ISOPLAST - engineering thermoplastic polyurethane resins
--- MAGNUM - ABS
--- PREVAIL - TPU/ABS alloy/blend
--- SABRE - polycarbonate/polyester alloy/blend
--- TYRIL - SA
--- STYRON - high-impact polystyrene
--- D.E.H. - epoxy Novolac resins
--- DERAKANE - vinyl ester resins
--- SPECTRIM - reaction moldable polyurethanes
--- VORANOL - polyols
--- ETHAFOAM - polyethylene foams
--- PULSE - polycarbonate/ABS alloy/blend

DSM ENGINEERING PLASTICS - EVANSVILLE, IN

DSM Engineering Plastics, U.S. subsidiary of NV DSM (The Netherlands), has several groups supplying plastics to the automotive industry. The Stanyl Marketing Center (Reading, PA) supplies STANYL nylon resins while The Polymer Corporation, also at Reading, PA provides NYLATRON nylon resins along with polyarylates. STANYL is a 4,6 nylon; NYLATRON nylons comprise the 6/6 and 6/12 moieties. Both STANYL and NYLATRON products are available in reinforced grades.

E.I DUPONT & COMPANY - WILMINGTON, DE

DuPont supplies the automotive industry with a variety of
engineering plastics which include:

--- DELRIN - polyacetals, including impact-modified version
--- RYNITE - thermoplastic polyesters (PET & PBT) reinforced with
      glass or mica/glass combinations
--- ARYLON - polyarylates
--- ZYTEL - 6, 6/6 and 6/12 nylons, often reinforced with glass
      (tradenamed GRZ) with super-tough nylons tradenamed ZYTEL ST
--- MINLON - combinations of nylon with minerals or mineral/glass
--- HYTREL - engineering TPEs (copolyesters)
--- VESPEL - polyamide
--- VITON - fluoroelastomer
--- BEXLOY - amorphous nylon
--- VAMAC - ethylene acrylic elastomer
--- TEFLON - PTFE
--- TEDLAR - PVDF
--- SURLYN - ionomer resin
--- SELAR - barrier resin
--- NORDEL - rubber
--- ALCRYN - melt-processable rubber

EASTMAN CHEMICAL - KINGSPORT, TN

Eastman Chemical, an Eastman Kodak subsidiary, provides several
plastic products to the automotive industry such as:

--- TENITE - polypropylenes and cellulosics
--- EKTAR - TPOs and filled polyesters and polyester blends

Eastman is promoting its polypropylenes as ABS replacements especially in instrument panels and inside door panels. Its TPOs are being used in fascias. The company's polypropylene plant is in Longview, TX.

ELF ATOCHEM NORTH AMERICA - PHILADELPHIA, PA

Elf Atochem North America is the U.S. subsidiary of Atochem (France), a division of Elf Aquitaine (France).

Elf Atochem's Plastic automotive product line includes:

--- RILSAN - nylon 11 and 12 - often used as powder coatings
--- PEBAX - nylon/polyether TPE
--- APPRYL - polypropylene
--- ORGALLOY PPO/nylon
--- PVC plastisols
--- CECA - phenolic resins

EMS-AMERICAN GRILON - SUMTER, SC

EMS-American Grilon, U.S. subsidiary of EMS-Chemie AG (Switzerland) produces engineering polymers at their plant in South Carolina.

The company's automotive plastic products comprise GRILON 6, 6/6 and 12 nylon; GRIVORY amorphous copolyamides, GRILAMID transparent nylons, and GRILPET thermoplastic polyesters (PET).

GE PLASTICS - PITTSFIELD, MA

GE Plastics, a division of General Electric, is a leading supplier of engineering polymers to the automotive industry. The company's product line consists of:

--- CYCOLAC - ABS
--- CYCOLOY - polycarbonate/ABS alloy/blend
--- GELOY - weatherable resins - polycarbonate/ASA
--- LEXAN - polycarbonates
--- LOMOD - engineering TPEs (copolyester)
--- NORYL - PPO/HIPS
--- PREVEX - PPE/HIPS
--- SUPEC - PPS
--- ULTEM - polyetherimides
--- VALOX - engineering polyesters
--- XENOY - polycarbonate/PBT alloy/blend

B.F. GOODRICH COMPANY - CLEVELAND, OH

B.F. Goodrich's Specialty Polymers & Chemicals Division supplies ESTANE TPUs to the automotive industry. These products can be injection- and blow-molded, extruded, melt coated, film laminated or solution coated. ESTANE was the first TPU, commercialized in 1960.

GENESIS POLYMERS - MARYSVILLE, MI

Genesis is a leading supplier of TPEs to the automotive industry. The company produces ADPRO - a polypropylene-based TPO in several grades: copolymer (reactor-produced), impact copolymer and performance homopolymer.

HOECHST-CELANESE CORP. - CHATHAM, NJ

Hoechst-Celanese, the U.S. subsidiary of Hoechst AG (Germany) is a major supplier of automotive plastics. Its product line consists of:

--- HOSTALEN - HDPE, polypropylene
--- CELCON - polyacetals
--- CELANESE - nylon 6, 6/6
--- CELANEX - thermoplastic polyesters (PBT)
--- FORTRON - PPS
--- VECTRA - liquid crystal polymers
--- RITEFLEX - TPEs (copolyesters) and alloys/blends
--- VANDAR - thermoplastic alloys
--- IMPET - thermoplastic polyesters (PET)
--- CELSTRAN - long-fiber-reinforced thermoplastics
--- DUREL - polyarylates
--- CELAZOLE - ultra high-performance materials

The company's U.S. headquarters is in Somerville, NJ; the Engineering Division at Chatham, NJ; Technical Center is located in Summit, NJ; and the Automotive Development in Auburn Hills, MI.

HUNTSMAN CHEMICAL CORP. - CHESAPEAKE, VA

Huntsman Chemical is an important supplier of polystyrene, polypropylene and specialty compounded polymers - although the company's forte is polystyrene. The high-impact version (HIPS) is alloyed/blended with several polymers (especially polyphenylene oxide) to produce important polymer alloys/blends with automotive applications.

Through a joint development program with GE Plastics, Huntsman Chemical introduced a new group of expandable thermoplastics tradenamed GECET. These materials are a blend of PPO, polystyrene impregnated with pentane (organic blowing agents). The automotive market will be targeted along with furniture and material handling applications. Specific automotive applications include: acoustic or thermal barriers, lightweight core materials, or as load-bearing functional members.

ICI Americas - Wilmington, DE

ICI Americas, a subsidiary of ICI, Ltd. (UK) is involved in the automotive plastics business via its Polyurethanes Group in West Deptford, NJ and ICI Fiberite/Composites in Tempe, AZ. ICI Polyurethanes also has a facility in Sterling Heights, MI.

ICI Americas automotive plastics products are made up of:

--- RIMline polyurethanes - elastomeric, rigid structural, glass-reinforced, low-density structural RIM
--- RUBIFLEX - polyurethane systems - flexible foams
--- FIBERITE advanced composites comprising phenolic, epoxy, and unsaturated polyesters

ICI's Advanced Material Group (Exton, PA) [recently sold to Kawaskai Steel] supplies VICTREX polyether sulfones (auto headlamp reflectors), MARANYL nylon 6/6, FLUON fluoropolymers (polytetrafluoroethylene) and POLYCOMP fluoropolymer composites.

181

MILES, INC. - PITTSBURGH, PA

Miles, the U.S. Bayer AG (Germany) subsidiary is another important automotive plastics supplier. The company's resins with automotive applications are:

--- APEC - high-heat polycarbonates
--- MAKROLON - polycarbonates
--- BAYBLEND - polycarbonate/ABS blend
--- MAKROBLEND - polycarbonate/polyester (PET) blends
--- TEDUR - PPS
--- DURETHAN - nylon 6
--- TEXIN - PC/TPU

Miles, a long time leader in polyurethanes, supplies the automotive industry with the following products:

--- flexible high-resilience (HR) foams for car seats, arm rests, and energy- and sound-absorbing applications.
--- semi-rigid foams (BAYFILL) for door and instrument panels; BAYFLEX for steering wheels, body-side moldings, and spoilers; energy-absorbing foam systems for bumpers and knee bolsters.
--- BAYFLEX RIM for modular window encapsulation.
--- SRIM for modular assembly
--- BAYFLEX reinforced reaction injection molding (RRIM) polyureas for body panels, door panels, fenders, etc.

MONSANTO COMPANY - ST.LOUIS, MO

Monsanto's Plastics Division provides the following plastics to the automotive industry:

--- LUSTRAN - ABS resins
--- VYDYNE - nylon
--- TRIAX - nylon/ABS alloys
--- CADON - SMA resins
--- CENTREX - weatherable ASA polymers

OWENS-CORNING FIBERGLASS - TOLEDO, OH

Owens-Corning supplies compression molded unsaturated polyesters (SMCs) to the automotive industry. The company's product line includes:

--- ATRYL resin - a one-component system delivering Class A
    surfaces for large panel and thick part applications
--- E-4297 - one component polyester resin system specifically
    for exterior automotive applications which are to be painted
--- E-980 - isophthalic resin system specially designed for
    structural grade SMC applications
--- DELTA P20 - a two-pack resin system designed to answer the
    need for improved high-performance and processing require-
    ments

183

PHILLIPS 66 COMPANY - HOUSTON, TX

Phillips 66, a division of Phillips Petroleum Company, produces
several plastic products for the automotive industry all under
the MARLEX tradename and include: polyethylene (mostly
high-density), polypropylene and RYTON polyphenylene sulfide.

QUANTUM CHEMICAL CORP. - CINCINNATI, OH

Quantum Chemical's USI Division produces several plastics used by
the automotive industry all under the PETROTHENE tradename which
include: HDPE, polypropylene, and TPOs (containing up to 50%
synthetic rubber and polypropylene).

REICHHOLD CHEMICAL CO. - RESEARCH TRIANGLE PARK, NC

Reichhold Chemical, a Dainippon Ink & Chemicals subsidiary is a
leading producer of unsaturated polyester resins tradenamed
POLYLITE. These phthalate-based materials are press molded into
SMCs and BMCs for automotive applications. Reichhold has
designed many of its resins for use in resin transfer molding
(RTM) and provides general purpose, structural and Class A
surface materials.

RESENOID ENGINEERING CORP. - SKOKIE, IL

Resenoid is a leading supplier of phenolic molding compounds
which are utilized in automotive under-the-hood applications such
as timing gears, brush holders, solenoid caps, starter motor
components, etc. One significant new automotive application is
the use of phenolics for molded throttle bodies and fuel rails.

The company's "1300 Series" are high-impact versions while the "2000 Series" are fabric reinforced phenolic molding compounds.

RHONE-POULENC, INC. - LOUISVILLE, KY

Rhone-Poulenc's Performance Resins & Coatings Division is part of Rhone-Poulenc, Inc., the U.S. subsidiary of Rhone-Poulenc (France).

The Division's plastics automotive products comprise: acrylics (headlight components) and EDI-REZ (base epoxies) for use in casting and SMC applications.

Rhone-Poulenc's worldwide business segments are made up of: organic and inorganic intermediates, specialty chemicals, fibers and polymer, health and agro.

A. SCHULMAN, INC. - AKRON, OH

A. Schulman is a leader compounder of plastic compounds with an Automotive Center located in Birmingham, MI. The company's automotive plastic product line includes:

--- POLYTROPE TPOs, typically blends of polypropylenes, EPDM or EP rubber - pre-colored versions used for rocker panels on 1992 Oldsmobile ACHIEVA.
--- POLYFAB ABS resins
--- POLYPUR TPUs (reinforced and alloyed)
--- POLYFORT heat-reinforced polyethylenes and polypropylenes
--- SCHULAMID - nylons

185

SOLVAY POLYMERS - HOUSTON, TX

Solvay Polymers is a Solvay America Company, the U.S. subsidiary of Solvay et Cie (Belgium). The company supplies FORTIFLEX HDPE and FORTILENE polypropylene to the automotive industry.

Solvay Group business segments were made up of the following: alkalis, peroxygens, plastics, processing and health.

TEXAPOL CORP. - BETHLEHEM, PA

Texapol Corp., a part of Cookson America (subsidiary of Cookson in the UK), provides processors with over one hundred engineering resins. These include thermoplastics such as nylon 6, 6/10 and 6/12, polyacetals, and polyesters (PBT). In addition, Texapol manufactures pre-colored matched resins and a line of natural, glass reinforced, impact modified, copolymer and heat-stabilized nylon products.

Many of these products, especially nylons, have interior automotive applications, e.g., seat belts, handles and door latches, window mechanisms, lighting.

THERMOFIL CORP. - BRIGHTON, MI

Thermofil Corp is a large plastics compounder serving the automotive industry similar to A. Schulman, Wilson-Fiberfill, etc. The company has a base of over 30 resins, over ten types of reinforcements plus a broad range of additives.

The company specializes in engineering thermoplastics and advertises its "leading edge technology" in composites especially with polypropylene, nylon and polycarbonates.

Printed and bound by CPI Group (UK) Ltd, Croydon, CR0 4YY

23/10/2024

01778230-0017